Why the Vietcong Fought

Recent Titles in
CONTRIBUTIONS IN POLITICAL SCIENCE
Series Editor: Bernard K. Johnpoll

Marx and the Proletariat: A Study in Social Theory
TIMOTHY McCARTHY

Privileged Communication and the Press: The Citizen's Right to Know Versus the Law's Right to Confidential News Source Evidence
MAURICE VAN GERPEN

Morris Hillquit: A Political History of an American Jewish Socialist
NORMA FAIN PRATT

The Establishment in Texas Politics: The Primitive Years, 1938-1957
GEORGE NORRIS GREEN

When Europe Speaks with One Voice: The External Relations of the European Community
PHILLIP TAYLOR

Project SEARCH: The Struggle for Control of Criminal Information in America
GORDON KARL ZENK

Unequal Americans: Practices and Politics of Intergroup Relations
JOHN SLAWSON

The Constitution of Silence: Essays on Generational Themes
MARVIN RINTALA

International Conflict in an American City: Boston's Irish, Italians, and Jews, 1935-1944
JOHN F. STACK, JR.

The Fall and Rise of the Pentagon: American Defense Policies in the 1970s
LAWRENCE J. KORB

Calling a Truce to Terror: The American Response to International Terrorism
ERNEST EVANS

Spain in the Twentieth-Century World: Essays on Spanish Diplomacy, 1898-1978
JAMES W. CORTADA

Why the Vietcong Fought

A Study of Motivation and Control in a Modern Army in Combat

WILLIAM DARRYL HENDERSON

Contributions in Political Science, Number 31

Greenwood Press
Westport, Connecticut • London, England

Library of Congress Cataloging in Publication Data

Henderson, William, 1922-
Why the Vietcong fought.

(Contributions in political science ; no. 31 ISSN 0147-1066)
Bibliography: p.
Includes index.
1. Vietnamese Conflict, 1961-1975. 2. Vietnam
(Democratic Republic, 1946-   ). Quân dôi–Organization.
3. Mặt trân dân tôc giai phống miên Nam Viêt Nam
I. Title. II. Series.
U742.H6   ~~355.02'09597~~   79-7062
ISBN 0-313-20708-9

Library of Congress Catalog Card Number: 79-7062
ISBN: 0-313-20708-9
ISSN: 0147-1066

First published in 1979

Greenwood Press, Inc.
51 Riverside Avenue, Westport, Connecticut 06880

Printed in the United States of America

10 9 8 7 6 5 4 3 2 1

Acknowledgments

The views expressed are my own and do not purport to reflect the positions or policies of the departments of the Army or Defense.

I would like to express particular gratitude to my professor, one of America's most knowledgeable individuals on Southeast Asia, Joseph J. Zasloff. Were it not for his assistance, encouragement, and gentle prodding, this book would not have been written. A great deal of credit also must go to others; any mistakes and misjudgments, however, are mine.

For my parents,
to whom I owe so much,
and for my sons,
for whom I expect so much

Contents

Figures

Tables

Preface

Two major armies fought in Vietnam between 1965 and 1972, the United States Army and the North Vietnamese Army (PLA).[1] One army endured, and the other did not. By all traditional methods of measuring military power, the final victory of U.S. forces should not have been in doubt. At the height of its combat power the United States Army in Vietnam consisted of 40 percent of all its combat-ready divisions supported by 50 percent of U.S. tactical air power and 33 percent of U.S. naval forces. Opposing this force was an army inferior in strength and vastly inferior in firepower and mobility capabilities. Never before had such massive U.S. firepower been concentrated on an opposing army in such a limited area for such an extended period of time. North Vietnamese casualties were heavy, reaching a peak of five for every United States and Allied casualty. Yet the North Vietnamese Army endured, maintained its cohesion, and remained on the battlefield when all others had retired.

Much has been written in explanation of the military outcome of the Vietnam War, particularly the factors prompting United States withdrawal. The willingness of the American people to withstand the enormous costs and deficiencies in the U.S. strategy of "graduated compellance" have been cited as major factors. Another factor of major significance was the weakened condition of the American army and in particular the resulting impact on the ability of the United States to achieve its goal of a favorable settlement in Vietnam. This has not been examined in great detail. What has been briefly discussed in the literature is the deterioration of morale and discipline within U.S. forces. Charles Moskos, for example, identifies a distinct progression of deterioration: "1965-1967, relatively high cohesion and morale; 1968-1969, a transition period of mixed cohesion and demoralization; and 1970-1972, widespread breakdowns in troop discipline."[2]

By the early 1970s, sharp contrasts between the United States Army and the PLA in morale, cohesion, and resulting combat power were readily apparent. "Combat refusals," fragging, a high incidence of open drug abuse, and other symptoms of widespread deterioration of morale and discipline became evident within U. S. forces in Vietnam. As the dysfunctional effects of these trends had their impact, the combat power of U.S. units eroded significantly.[3] Aggressive search-and-destroy missions were no longer ordered. It was thought prudent to assume a defensive posture with only local patrolling in the vicinity of U.S. bases and landing zones. The primary motivations of many individual American soldiers during this period were seen as follows:

> Most observers agree that discipline and morale were at rock bottom in the Army in Vietnam in 1971. It was even said that many of the replacements arriving were volunteers for repeat tours who had been attracted by the additional combat pay, the availability of cheap drugs, or the slacker discipline. Units continued to function, even to fight when necessary for self-defense, largely through an unspoken compact between "lifers" and "grunts." The rules were simple—stay alive, finish your year, and go home.[4]

In sharp contrast to this picture of organizational decay was the steadfast cohesion and endurance of the PLA. Although brief comments on current explanations for the "collapse of the United States Army in Vietnam" are made in the preface, the major effort of this study is to examine PLA organization, policies, and motivation to learn how the PLA endured and survived the most continuous and intense firepower ever directed against any army.

It will be shown that the PLA soldier was not an unbeatable guerrilla fighter motivated by a mystically potent combination of communism and nationalism. Instead, the PLA soldier was much like the U.S. soldier who suffered under the stress and hardship of combat. Fear, lack of sleep, rain, heat, and fatigue affected the PLA much the same as they did the Americans; however, a vital difference between the two armies was their internal organization and procedures.

The United States Army experienced an extremely complex phenomenon in Vietnam. Major factors such as U.S. domestic politics, the "greening of America," racial friction, and international politics as well as significant organizational characteristics and policies within the army itself combined to exert significant pressure upon U.S. troops in Vietnam. Because the army did nothing to prepare the individual soldier and small unit for the effects of these factors a distinct and ever-increasing deterioration of cohesion and morale occurred from 1965 through 1972.

It should be stated from the first that not much reliable evidence is available concerning the motivation and morale of the American soldier in Vietnam. Other than very subjective media reports, there are few sources of information suitable as primary sources for in-depth, academic research. I am aware of only one significant, but limited, attempt to collect information about the American soldier in Vietnam which would be suitable as a basis for generalizing about discipline and cohesion in American units. This was done by Charles Moskos in 1965 and again in 1967. These two separate data-gathering expeditions lasted one and two weeks, respectively, and focused on a squad-sized unit each time.[5]

To my knowledge, there were no other significant programs or attempts to collect a large number of interviews designed to shed light on the motivation, morale, and cohesion of soldiers within U.S. units.[6] This regrettable lack of primary-source interviews and documents suitable for valid generalizations concerning the American soldier in Vietnam did not apply in the case of the PLA. Hundreds of interviews of PLA soldiers are available.[7] Tremendous sums were spent to collect and analyze the data uncovered in these interviews. Thus one is able to comment quite extensively about discipline and cohesion within the PLA but not about the United States Army in Vietnam. However, Moskos and others offer some observations which are in general accordance with antecedent literature on this subject.[8]

Singificant factors frequently cited as leading to the breakdown in discipline and cohesion in the United States Army in Vietnam are: the one-year combat tour; other similar personnel policies within the United States Army in Vietnam; and broad societal factors such as racial strife, drug usage, and generational differences. There is little primary-source evidence that would support such conclusions. However, on the basis of available literature and studies, one may offer a working hypothesis, realizing there is little direct evidence to support any hypothesis.

It is proposed that cohesion and morale were high upon the initial commitment of U.S. units to Vietnam, especially in those units which trained together in America, moved to Vietnam as a unit, and entered combat together. However, once in-country rotation of officers and enlisted men was effected and the extremely deleterious effects of the one-year combat tour and other similar policies had their impact, the cohesive fabric of U.S. units began to unravel.

Small group norms, so important to cohesion and discipline within units, became less significant in determining the daily behavior of soldiers in these units. As personnel turbulence increased, units comprised essentially of strangers developed. Friendships, working relationships, and mutual expectations of individual and unit behavior became superficial. In other words, strong primary group relationships so necessary for small-unit cohesion in combat gradually weakened to the extent that they were not, in many cases, the overriding determinant of the individual soldier's behavior. It is proposed that the condition of military anomie which developed in many units provided fertile ground for new and deviant group norms to emerge. In many cases these norms were not congruent with organizational goals and the new patterns of behavior that emerged from such forces as the drug, cultural, and racial friction of the times often resulted in overt action against the military (for example, combat refusals, fragging, etc.).

The emergence of such a situation is not difficult to understand. The relatively firm unit cohesion within the army in Vietnam in 1965 gradually gave way to unit environments in which strangers were thrown together for a few months in an atmosphere of organizational impermanence and uncertainty. Such units were unable to provide the basic organizational environment necessary for combat units to endure cohesively.

Thus the combined effects of the many expedient and often self-inflicted dysfunctional policies adopted in Vietnam did not engender units capable of sustaining the strong primary groups necessary for cohesive and disciplined units. The United States Army in Vietnam lost sight of the basic truth that combat is a group action, requiring the utmost in mutual cooperation; in "good" units, cooperation is motivated by an intense need on the part of the individual soldier to fulfill recognized obligations toward his fellow soldiers.

By the early 1970s cohesion had unraveled in so many U.S. units in Vietnam and deviant norms were gaining sway to such an extent that one of the highest ranking and most well-known generals in the army stated:

> I will be damned if I will permit the United States Army, its institutions, its doctrine and its traditions to be destroyed just to win this lousy war.[9]

The United States Army must learn from its experiences in Vietnam. Like any unpleasant experience, the tendency since our withdrawal has been to avoid and forget Vietnam, at least at the institutional level. Preoccupation with the army's NATO commitment and related priorities preclude the dedication of limited resources to examining in any comprehensive way the lessons of Vietnam. By default this has been left to others. Unfortunately, some critics have not been restrained by responsible scholarship and accurate reporting.

In this regard the most recent and perhaps most potentially harmful commentary on the United States Army's experience in Vietnam was made in a recent book by Gabriel and Savage.[10] It abounds with egregious assertions and unbalanced conclusions which could provide the basis for major misconceptions among unwary readers about what actually happened to the army in Vietnam and, more importantly, why it occurred.

Gabriel and Savage propose three interrelated hypotheses: (1) the army underwent a progressive disintegration of cohesion in small units which culminated in the early 1970s in Vietnam; (2) the "disintegrative" process operated independently of sociopolitical factors in the larger American society; and (3) the disintegration of the army and loss of small-unit cohesion "is directly related to the loss of officer professionalism. . . ."[11]

In support of their hypotheses, the authors pursue a somewhat reductionist and linear logic of cause and effect. A bare outline of their logic progresses as follows: The United States Army Officer Corps has adopted a managerial ethos in which a major goal is individual career advancement. As a result, the officer corps has become, according to the authors, careerist, ill-trained, and corrupt. This lapse of "traditional" leadership led to the disintegration of primary group cohesiveness which in turn led the soldiers to such antimilitary acts as desertion, "fraggings," and combat refusals, eventually bringing about the disintegration of the United States Army in Vietnam.

The authors claim attempted reforms have been ineffective and the basic malaise still exists within the army; because of it, the United States faces "rapid defeat" in the event of another war.

Such questionable scholarship and emotive indictments of the officer corps and, more importantly, the authors' prognosis of future defeat because of malfeasance and careerism within the corps deserves detailed critique.

Crisis in Command has serious methodological flaws. One that is not only major but is likely to pass unnoticed because of its enormity is the fact that Gabriel and Savage have no primary sources of evidence in support of their basic hypotheses. Nowhere do they show evidence that alleged malfeasance, careerism, etc., within the officer corps had a significant and major impact on the American soldier and caused disintegration within units in the army in Vietnam.

The authors' main source is the United States Army War College's *Study on Military Professionalism*, which itself has serious methodological flaws and, furthermore, was neither designed nor intended for the use chosen by Gabriel and Savage.[12] The study does not contain one interview of United States enlisted soldiers in Vietnam and the authors cite no other sources which directly support their proposition that American soldiers stopped following their leaders because of incompetence and corruption within the officer corps. The interesting portion of the study is the intraprofessional commentary by junior officers about more senior officers. It contains nothing about American enlisted men or their attitudes or perceptions about officers' performance in combat. To substantiate their assertions, the authors would have to demonstrate widespread perceptions on the part of soldiers that they lost faith in the ability and purpose of their officers. This was not done because such evidence was and is not available. Because the authors cite no other primary sources, one must conclude they have made an a priori assumption that officer unprofessionalism was the root cause of the "collapse" in Vietnam.

Because neither Gabriel nor Savage served in Vietnam, to generalize as they have done would require sources similar to those cited in several well-known and respected studies of different armies. Studies of the Wehrmacht in World War II by Shils and Janowitz, of the United States Army in World War II by Stouffer et al., of the Chinese in Korea by George, of the Americans in Korea by Little, of the Americans in Vietnam by Moskos, and of the North Vietnamese PLA (see Introduction) all rely on detailed and, in most cases, extensive interviews with the soldiers under discussion. Unfortunately, this is not the case with *Crisis in Command* and the authors' unsupported assertions lead them far beyond the bounds of academic responsibility.

In addition to this major flaw in methodology, the book contains numerous other misstatements and assertions central to Gabriel's and Savage's case. At the outset it should be stated that I agree with the authors' proposition that disintegration of small-unit cohesion was central to the dismal performance of the United States Army during its latter years in Vietnam. However, I take strong exception to their assigning causal responsibility for the loss of cohesion to officer unprofessionalism and worse. The weaknesses in such an approach are outlined below.

Gabriel and Savage charge that "disintegration"was "clearly associated with the large numerical expansion of the Officer Corps to levels previously unknown."[13] However, no reliable evidence is cited supporting their assertion that officer strengths were a major force in causing small-unit disintegration. It is a great leap from citing a table of worldwide officer strengths to demonstrating that troops in front-line units in Vietnam were aware historically of officer-enlisted men ratios and, further, that this had a direct and negative impact on unit cohesion. Moreover, the officer structure of front-line units in Vietnam does not support this assertion as the officer structure in platoons, companies, and battalions has changed little since World War II. Neither interviews nor survey data of front-line troops of any type is offered in support of this assertion. Instead, the authors offer a historical review of worldwide officer-enlisted men ratios from 1918 through World War II to 1972. This review reflects that the officer portion of the ratio in 1972 was double that required in World War II. On the basis of this "evidence" Gabriel and Savage come to the amazing conclusion that disintegration was "clearly associated with the large numerical expansion of the officer corps to levels previously unknown."[14] Such analysis is clearly unacceptable. No effort was made to determine the combat troops' perceptions of officer strength and assess its impact on "disintegration." No examination was made of officer-enlisted men ratios in front-line units. As already noted, if such an examination had been made, the authors would have found little change from World War II. In sum, much inconclusive consideration was given to worldwide officer strengths without demonstrating how it impacted on front-line cohesion in Vietnam (if it did).

The authors imply Vietnam was the major cause for the increased officer-enlisted men ratios when in fact it was not. No consideration was given to the radically different international and strategic situation the United States has had to deal with since the postwar emergence of the USSR in 1948 as a major military competitor and how resulting alliances and supporting military organizations became major factors in accounting for increased officer strengths over those required in World War II. In short, the many alliances and preparations for instant and worldwide nuclear and conventional war greatly increased reliance on advanced technology; resulting requirements for significantly more officers to spend more time in more schools are just a few factors that were not considered by the authors in explanation of increased officer-enlisted men ratios since World War II.

In a related assertion the authors charge that increased officer strength associated with Vietnam caused "the quality of the average officer. . .to suffer."[15] Again, the evidence offered is questionable. The authors state that "with the withdrawal of so many colleges and universities from the ROTC system. . . officer quality had to decline."[16] The implication is that the decline in universities participating in ROTC led to a decline in the number of officers normally received from the ROTC program. By focusing on the fact that some universities discontinued ROTC, the authors failed to note that the number of officers commissioned through ROTC actually increased steadily during the Vietnam War.[17]

In developing this point further, Gabriel and Savage state: ". . . no society, especially a mass society, can ever produce more than a very limited number of military leaders who can be expected to lead men directly and to do well in battle."[18] They then show a historical statistical summary of officer-enlisted men ratios and conclude with the statement that "compared to World War II the number of officers had increased by almost 100 percent."[19] The implications of their argument are obvious, yet a closer look at the actual numbers reveals a complete lack of support for their position and in fact tends to refute their point.

The authors cite the United States Army during World War II as a model of small-unit cohesion and combat performance.[20] However, as noted, they state a society has a limited ability to produce combat leaders and that rapid expansion of the Officer Corps will lead to decreased officer quality. Yet, using their own table, it is readily apparent that officer strength during World War II was almost three times that of the Vietnam era.[21] Thus, according to the authors' basic proposition, the quality of officers serving in Vietnam should have been significantly better than that of those in World War II. However, by some quirk in logic they claim the opposite. A detailed look at the actual numbers of officers involved in the Vietnam buildup reveals only a nominal increase of 50,000-60,000 officers. One would hardly expect that such a small increase would be a strain on the ability of the United States to produce combat leaders, especially since the army produced a high of 481,000 such leaders during World War II (not including the Army Air Corps) while the nation was called upon to provide only 171,000 officers during the height of Vietnam (including army aviators).[22]

In another unconvincing assertion, Gabriel and Savage state that "one major evidence of disintegration frequently cited is the desertion rate";[23] they go on to say their data shows that "Vietnam desertion rates exceeded those of World War II by over ten per thousand at their height. . . ."[24] Such analysis is inept if not purposely misleading. In Table 1 the authors imply the desertion rate in Vietnam ranged, in the latter years, from approximately fifty-two per thousand soldiers to a high of seventy-three per thousand. This was not the case. Desertion in Vietnam remained at about five per thousand even during the peak desertion period in the rest of the army. The desertion rates cited by the authors are for the entire army, worldwide. This is extremely misleading and appears to be the result of sloppy analysis. For example, a careful reading of their Table 2 reveals the number of desertions in 1972 (44,643) was almost double the number of troops in Vietnam (25,200). In their unbalanced analysis, the authors not only fail to support their point, they fail to link desertion to their basic proposition. In other words, if desertion was a function of poor leadership and was a major cause of disintegration in U.S. forces in Vietnam, the authors fail to explain why the in-country desertion rate never exceeded five per thousand, well below the rate in the rest of the army. They may belatedly argue that these soldiers had no opportunity to desert in Vietnam and were deserting on the way to Vietnam or after their return. This thesis also is not supported by fact. Data from the President's Clemency Program indicates the following about deserters during this period:

> Their service careers tended to be short, most (75%) served two years or less, few (19%) saw service in Vietnam, and fewer yet (1%) deserted from combat. Many (44%) had been previously AWOL. Occupational shifts and reduced rank also pointed to histories of trouble in the service among these men.
>
> *Their reasons for leaving were generally unassociated with the war.* Most (50%) stated that they had left because of personal/family/or financial problems (the same reason given by most deserters during the last two wars as well). Only 14% of the men mentioned Vietnam as in any way responsible for their decision to leave the Army [emphasis mine].[25]

Another outrageous assertion is that U.S. officers shirked the dangers of combat in Vietnam. Gabriel and Savage state:

> Relative to their numbers and particularly the higher ranks, American Army Officers did not share the burden of death that they asked of the men they sent to fight. Indeed, the number of enlisted men upon whom the burden of death fell were themselves but a small part of the total forces. In 1968, at the height of the buildup, fewer than 80,000 combat forces could be put into the field against a maximum ground troop strength in Vietnam of 543,000.[26]

What the first sentence has to do with the remainder of the quotation and the authors' main point is unexplained. The indiscriminate use of such data could mislead the unwary reader and add unwarranted credibility to the authors' assertion that officers in Vietnam did not share the dangers of combat as they had in past wars and thus officer "immunity from harm" was a major factor in the disintegration of the United States Army in Vietnam. As it is, the authors' own data does not support their assertion. An examination of Table 7 reveals that the ratio of officer-enlisted men battle deaths in Vietnam for:

Colonel	were lower than World War II but higher than Korea.
Lt. Col.	were higher than both World War II and Korea.
Major	were higher than both World War II and Korea.
Captain	were higher than both World War II and Korea.
1st Lt.	were higher than both World War II and Korea.
WO	were higher than both World War II and Korea.
2d Lt.	were lower than World War II and the same as Korea.[27]

This data hardly supports the authors' point. The primary reason the 2d Lieutenant ratio is lower is that it was army policy not to send inexperienced 2d Lieutenants to Vietnam immediately upon being commissioned. Those that did go to Vietnam were relatively senior and were soon promoted to 1st Lieutenant.

Gabriel and Savage's comments to the effect that warrant officers in Vietnam died alone with no impact on their unit and therefore should not be considered in comparing World War II officer deaths with those in Vietnam reflect both authors' unfamiliarity with the Vietnam War.[28] Had either served in Vietnam he would have realized that front-line aviation units were among the most cohesive U.S. units in Vietnam and the warrant officer pilot was central to the creation and maintenance of such unit esprit. Presumably, cohesion and disintegration in front-line U.S. units is the focus of Gabriel and Savage's effort, yet it is curious that they dismiss its importance among those officers and units which carried a very significant share of the combat mission in Vietnam.

An additional comment that indicates a profound misunderstanding about the nature of the war in Vietnam is the authors' statement that the greatest number of sacrifices (deaths) fell upon the "lower-ranking officers and enlisted men" with direct implication that this was so primarily because the more senior officers shirked their traditional duties by avoiding their leadership responsibilities and engaged in a practice of "death management" from a distance whereby lower-ranking men were "managed to their death" by officers who were largely safe and secure in rear areas, etc.[29] As evidence, Gabriel and Savage point to past wars and armies in which high-ranking officers (including general officers) suffered casualty rates much higher than those suffered by colonels and generals in Vietnam.[30] This is fallacious. No consideration was given to the fact that Vietnam was a war fought primarily at the platoon and company level. There were no major land battles, tank battles, aerial dog fights, opposing bomber forces, and artillery duels similar to those of World War II. The simple fact is that senior officer casualties were lower in Vietnam because of the nature of the war and the proper role of senior officers in such a war. There was no shirking of duty or avoidance of "traditional" roles. Vietnam was an unconventional war that did not accommodate the authors' limited notion of the officer's "traditional" role.[31] Had either served in Vietnam the inappropriate nature of these and other assertions would have been apparent.

In laying the collapse and "discohesion" of the United States Army in Vietnam entirely at the feet of the Officer Corps, Gabriel and Savage make another extraordinary assertion by stating:

> To a significant degree, the disintegrative process operated independently of sociopolitical factors in the larger American Society. . . .we do not agree with popular notions that indicators of military disintegration are the direct result of factors generated by the external society.[32]

This assertion is extraordinary in that no evidence is offered in support—just the statement in the form of a conclusion. Such an unsupported statement is simply unbelievable, especially since it is made in the face of numerous suggestions by respected analysts that there were indeed significant linkages between events in American society and similar events within the United States Army in Vietnam. Just the fact that the rise of incidents reflecting "disintegration" in Vietnam closely paralleled the rapid increase of generational, racial, and drug phenomena in the United States should make one extremely cautious about dismissing any significant linkages between them.

In sum, it is my contention that not much can be said with great authority about why the United States Army evolved as it did in Vietnam. There have been various explanations, some more convincing than others, but none has been definitive. Certainly Gabriel and Savage are not, and it is likely that no definitive studies will emerge given the lack of reliable data.

Compared to the performance of the United States Army in Vietnam, the PLA was a wonder of organizational effectiveness. The remainder of this study examines the organization and associated policies of the PLA and concludes that they were major factors in explaining PLA cohesion and endurance.

NOTES

1. PLA=People's Liberation Army. The term *PLA* also includes the Vietcong (VC). See note 5 of Introduction for a more complete explanation of the term PLA.
2. Charles C. Moskos, Jr., "The American Combat Soldier in Vietnam," *Journal of Social Issues*, vol. 31, no. 4, 1975, pp. 25-37.
3. William L. Hauser, *America's Army in Crisis* (Baltimore: Johns Hopkins Press, 1973), pp. 73-130; Stuart H. Loory, *The Defeated: Inside America's Military Machine* (New York: Random House, 1973), pp. 70-73.
4. Hauser, *America's Army*, p. 104.
5. Charles C. Moskos, Jr., *The American Enlisted Man* (New York: Russell Sage Foundation, 1970), p. 136.
6. There was another field trip to Vietnam conducted by members of the U.S. Military Academy faculty during the summer of 1970 for the purpose of determining soldiers' morale and attitudes toward their immediate superiors. These interviews were for pedagogical use and have never been used to generalize about U.S. troop attitudes in Vietnam.
7. See Appendix A for an in-depth discussion of these interviews and research projects.
8. See Introduction for a review of this literature. Moskos differs to the degree that he maintains the unique policies of the United States Army in Vietnam undercut the cohesive effects of the primary group and produced an "essentially individualistic soldier in Vietnam." See Moskos, *American Enlisted Man*, pp. 144-46.
9. Ward Just, *Military Men* (New York: Avon Books, 1972), p. 189.
10. Richard A. Gabriel and Paul L. Savage, *Crisis in Command: Mismanagement in the Army* (New York: Hill and Wang, 1978).
11. Gabriel and Savage, *Crisis in Command*, pp. 8-12, 31, 51, 59, 63.
12. *Study on Military Professionalism* (Carlisle Barracks, Pa.: U.S. Army War College, 1970).

13. Gabriel and Savage, *Crisis in Command*, pp. 61-62; see also p. 9.
14. Ibid., p. 62.
15. Ibid.
16. Ibid.
17. T. O. Jacobs, unpublished, undated information paper (Alexandria, Va.: Department of the Army Individual Training and Performance Research Laboratory, 1978).
18. Gabriel and Savage, *Crisis in Command*, p. 53.
19. Ibid., p. 62.
20. Ibid., pp. 31-41.
21. Ibid., Table 5.
22. Ibid., p. 185.
23. Ibid., p. 42.
24. Ibid., p. 43.
25. Quote from D. Bruce Bell and Thomas Houston, Jr., *The Vietnam Era Deserter* (Alexandria, Va.: Army Research Institute Report 76-6, 1976), p. iii.
26. Gabriel and Savage, *Crisis in Command*, p. 63.
27. Ibid., p. 187.
28. Ibid., pp. 66-67.
29. Ibid., pp. 54, 63-67.
30. Ibid., p. 39. In addition to the United States Army in World War II, the authors cite the Wehrmacht and the recent history of the French and British armies. In this regard it is interesting to note that they cite the battle of Verdun as an example of officers performing correctly in their traditional role. This is curious because Verdun is considered by most military historians to be a prime example of officer failure to recognize the changing nature of warfare, thus resulting in massive and unnecessary loss of life.
31. Ibid., p. 187. When the overall rate of officers killed in Vietnam (1.82 per thousand men) is compared to the overall rate of enlisted men killed in Vietnam (1.87 per thousand men), it is readily obvious that there is no significant difference.
32. Ibid., pp. 31-33.

Introduction

The question of why the Vietcong fought so well against such tremendous odds has often been asked. Many familiar arguments have been presented in explanation. Vietnamese nationalism, resistance to "imperialism," frustration due to economic and social deprivation, and the effects of an aggressive communist ideology are only some that have been suggested.

One approach to this question is through an examination of motivation and control.[1] These concepts are used to show how an organization such as an army manipulates its members in order to accomplish organizational goals. While they may be linked to the considerations already mentioned they focus specifically upon the behavior of the individual soldier and the small groups to which he belongs.

The nature of modern warfare makes this approach profitable. In examining motivation within modern military organizations—that is, essentially asking why men fight—Morris Janowitz concludes that modern warfare has required a continuing shift from external control schemes based on "domination" toward control devices based on some type of normative "manipulation."[2] Modern firepower no longer permits the attack of mass formations under the watchful eye of noncommissioned officers. In Professor Janowitz's words, the combat soldier no longer resembles the model "of Max Weber's ideal bureaucrat following rigid rules and regulations."[3] The emphasis on strict discipline, drill, and rote training has steadily declined. While still important, but to a lesser degree, these factors tend to produce a "routinized" soldier. Modern weapons and tactics have made the direct control of troops in combat exceedingly difficult, if not impossible, even for a platoon leader or company commander. The dispersion, confusion, chance, and danger which characterized battlefields in Indochina make obsolete even the formations and techniques of control relied upon in World War II for the direction of firepower and the maneuver of troops. Along with the decline of classical battle formations, emphasis on strict orders, discipline, drill, and rote training there has been a corresponding shift downward in the locus of control in combat, and increasing attempts to use organization to motivate and control the individual soldier and gain some measure of his loyalty. This is especially true in guerrilla or counterinsurgency-type armies where there must be constant improvisation and initiative by individuals and small units.

This study investigates the nature of the relationships which existed between the fighting men in PLA combat units and the Communist party organization for which they fought. The importance of the "human element," borrowed by General Giap from the Chinese PLA concept of "man-over-weapons," is explained in the following quotation from Mao Tse-tung:

> In all armies, obedience of the subordinates to their superiors must be exacted. This is true in the case of guerrilla discipline, but the basis for guerrilla discipline must be the individual conscience. With guerrillas, a discipline of coercion is ineffective. . .discipline must be self-imposed, because only when it is, is the soldier able to understand completely why he fights and how he must obey. This type of discipline becomes a tower of strength within the army, and it is the only type that can truly harmonize the relationship that exists between officers and soldiers.[4]

Mao points out that under the conditions of the modern battlefield, where the individual is largely alone except for two or three close comrades, the formal organization has no direct means of controlling his behavior. In such a situation the importance of the small combat unit to which the soldier belongs cannot be overemphasized. The small group which develops around intimate face-to-face relationships in a combat environment is the immediate determinant of much of the individual soldier's behavior. Thus, it becomes the task of the organization, in this case the party, to dominate the small fighting group through its cadre.

This study is significant for several reasons. As the most successful application of the Chinese PLA model (George 1967) outside of China, the Vietnamese PLA experience attracted much attention as it developed and refined a distinctly Vietnamese army.[5] Comparisons of the PLA with the Chinese model should be interesting to students of military motivation and control. For example, there is some evidence that the Vietnamese PLA relied less on overt control and surveillance techniques than did the Chinese PLA in Korea when it met severe setbacks similar to the Vietnamese PLA experience from 1965 to 1967.

When the Vietnamese communists decided to achieve by force what could not be achieved peacefully they required a disciplined army which the party could count on as a reliable and predictable instrument. In creating such an army the party found certain constraints with which every modern army must deal. Much can be learned from the PLA experience.

A different area where this study may be useful has to do with any question concerning the possibility of future insurgencies in Southeast Asia or elsewhere based on the Vietcong/PLA model. Anyone investigating the likelihood of "another Vietnam" in Thailand, Burma, or even Africa or Latin America will want to compare the organization, the cadre, and the resources of the insurgents to similar factors in the Vietnamese PLA. Such comparisons would help evaluate future predictions of "domino theories." The quality of armed forces or potential armed forces available to an insurgent organization is an important factor in assessing the probability of future "falling dominoes." For example, the recent insurgent victories in Southeast Asia have stopped short of Thailand. The emergence of this new "frontier" is in part explained by the lack of a well-organized insurgent armed force within Thailand. Whether Thailand becomes a "domino" will be determined in large part by insurgent success in creating a Thai organization similar to the PLA.

This study may also help in dealing with the mysticism inherent in the quasi-myth of the "unbeatable peasant guerrilla army." Such an army is often presumed to be composed of indefatigable fighters who are motivated by a vague "rice paddy" type of populism and pose no control or motivation problems for their leadership. However, the PLA soldier will be shown to be much like any other soldier exposed to danger and hardship. Fear, cowardice, anxiety, privation, brutality, sickness, and evasion of duty were often evident in the behavior of PLA soldiers. This study shows how the cadre worked within the PLA organization to negate these factors.

An explanation of why the individual soldier fought for PLA objectives is to be found in the interaction between the party organization, the cadre, and the primary group to which the individual soldier belonged. From the viewpoint of the party the challenge was to ensure that the accepted informal rules of conduct which emerge in a small face-to-face combat group and govern individual behavior were congruent with PLA goals. This was accomplished for the party by the cadre-leader who attempted to dominate the small group in all respects and thereby ensure that its goals were those of the party.

Many investigations of why men fight have focused on the concepts of "morale" or "esprit de corps" and have discussed individual and unit performance in combat in terms of courage, discipline, enthusiasm, and willingness to endure hardship. However, such research does not explain the factors involved in the endurance of the PLA as a professional army.

According to Janowitz (1964), "even in the smallest unit there is an 'iron framework' of organization which serves as a basis of social control. The single concept of military morale must give way, therefore, to a theory of organizational behavior in which an array of sociological concepts is employed" (Janowitz and Little 1965; George 1967).

The literature on military motivation suggests a number of explanations for man's behavior in combat which are useful in investigating the PLA. These approaches treat the primary group and its relationship to the organization in explaining combat behavior. Beginning with Shils and Janowitz in their study of cohesion and disintegration in the Wehrmacht, small group cohesion, interaction within the group, and organization have been increasingly emphasized.

By the term *primary group* most investigators refer to the concept of *Gemeinschaft* (small intimate community relationships). More specifically, primary groups have been conceptualized as being:

> . . .characterized by intimate face-to-face association and co-operation. *They are primary in several senses, but chiefly in that they are fundamental in forming the social natures and ideals of the individual.* The result of intimate association. . . is a certain fusion of individualities into a common whole, so that one's very self, for many purposes at least, is the common life and purpose of the group. Perhaps the simplest way of describing this wholeness is by saying it is a "we" [emphasis mine].[6]

Research indicates that the soldier is strongly bound to the primary group as long as it is capable of satisfying his major physiological and social needs. Shils and Janowitz reported that as long as the Wehrmacht soldier had the necessary resources and as long as the primary group met his essential personal needs he was "bound by the expectations and demands of its other members." Molnar (1965) cites similar evidence discussing soldiers bound to some degree by social role and status patterns common to a primary group. Shils and Janowitz state:

> It appears that a soldier's ability to resist is a function of the capacity of his immediate primary group (his squad or section) to avoid social disintegration. When the individual's immediate group, and its supporting formations, met his basic organic needs, offered him affection and esteem from both officers and comrades, supplied him with a sense of power and adequately regulated his relations with authority, the element of self-concern in battle, which would lead to disruption of the effective functioning of his primary group was minimized.[7]

Shils and Janowitz support this basic hypothesis by observing that:

> For the ordinary German soldier the decisive fact was that he was a member of a squad or section which maintained its structural integrity and which coincided roughly with the social unit which satisfied some of his major primary needs. He was likely to go on fighting, provided he had the necessary weapons, as long as the group possessed leadership with which he could identify himself, and as long as he gave affection to and received affection from the other members of his squad and platoon. In other words, as long as he felt himself to be a member of his primary group and therefore bound by the expectations and demands of its other members, his soldierly achievement was likely to be good.[8]

Additional factors also impact upon the cohesiveness of the primary group and its influence on the behavior of the soldier. Many investigators have pointed out that the concept of the primary group takes on an added sharpness under combat conditions. In considering the primary group as a dependent variable, the mere fact that a combat situation entails an increase in solidarity in response to an external threat is a phenomenon which has been verified many times. When a threat and the responsibilities for coping with it are shared, an increase in group solidarity and a reduction of internal group conflict usually occur. Observers of men in combat have called attention "again and again to the fact that the most significant persons for the combat soldier are the men who fight by his side and share with him the ordeal of trying to survive."[9] S.L.A. Marshall, an observer of men in combat in numerous wars, states: "I hold it to be one of the simplest truths of war that the thing which enables an infantry soldier to keep going with his weapon is the near presence or the presumed presence of a comrade."[10]

Another variable which seems to increase primary group cohesion in combat is the soldier's calculation of his chances for escape from the threatening situation. If he is bound to the primary group by isolation from surrounding groups, by anxiety-producing doubts about his ability to leave his unit successfully, and other such ambiguities, he sees his best chance of survival as resting with one or two "buddies" or with the other members of his primary group (Little 1964).

Other factors influencing primary group cohesiveness are the past social experiences of the members. Common religion, race, ethnic group, social class, age, region, and history appear to contribute to the communications necessary for intimate interpersonal relationships common to a primary group (Janowitz and Little 1965; Shils and Janowitz 1948; Emerson 1967; Kohn 1932; George 1967).

Another influence shaping primary group solidarity is the member's commitment to his sociopolitical system, ideology, secondary group symbols, and causes such as common awareness and resentment of the "nation's" colonial history (George 1967). Indoctrination induces commitment to secondary symbols by establishing preconditions for primary group cohesion. Indoctrination themes generally stress the legitimacy of war aims and justify fighting for such aims (George 1967). While recognizing the impact of secondary groups on the individual soldier, Shils and Janowitz maintain their influence is slight compared to that of the primary group. They quote a German soldier as saying:

> The company [military unit] is the only truly existent community. This community allows neither time nor rest for a personal life. It forces us into its circle, for life is at stake. Obviously compromises must be made and claims be surrendered. . . . Therefore the idea of fighting, living, and dying for the fatherland, for the cultural possessions of the fatherland, is but a relatively distant thought. At least it does not play a great role in the practical motivations of the individual.[11]

The honor and romanticism involved in fighting a war often appeal to the young soldier who experiences the need for asserting manliness or toughness. The coincidence of these personal needs with similar group norms and military codes also serves to reinforce group solidarity (Shils and Janowitz 1948; Stouffer et al. 1949, Moskos 1970).

This discussion has emphasized the influence of the primary group in shaping the behavior of the soldier. However, a significant question remains. Will the primary group produce behavior by the soldier that is congruent with the goals of the organization? Many investigators have noted that the primary group cohesiveness which emerges in the small combat unit can militate either for or against the goals of the formal military organization (Etzioni 1961; Janowitz and Little 1965; George 1967). For example, Janowitz and Little, in discussing problems of segregated "Negro" units during World War II, state:

> Primary groups can be highly cohesive and yet impede the goals of military organizations. Cohesive primary groups contribute to organizational effectiveness only when the standards of behavior they enforce are articulated with the requirements of formal authority.[12]

Still other investigators have found small group behavior in combat situations that is deviant from the organization's point of view (Shils and Janowitz 1948; George 1967; Little 1964). Shils and Janowitz in their investigation of cohesion and disintegration in the Wehrmacht found that units that surrendered as a group were led by "soft-core," non-Nazi comrades, to whom organizational goals were relatively unimportant.[13]

The performance of the group in meeting organizational goals is largely dependent upon the effectiveness of the leader. Research suggests that a capable leader can manipulate primary group members through a wide range of organization mechanisms, psychological techniques, and indoctrination themes in order to shape primary group norms and attitudes that are compatible with organizational objectives. He can accomplish this because he has been accepted as the natural leader of the small group. Men who fight modern wars must be convinced that their leaders have their welfare in mind, and leaders must continually demonstrate expertise and set the example in adhering to group norms before men will follow them.[14]

Primary group behavior, whether deviant or desirable from the organization's point of view, is the result of norms formed by primary group interaction. The primary group is therefore a major factor in explaining man's behavior (positive or negative) in combat.

It is suggested that the Vietnamese communist leadership was acutely aware of these considerations and extremely wary of any group it did not intimately control. It is further suggested that the party leadership realized that such groups were instrumental in controlling man's behavior in combat and when properly motivated would achieve party military objectives.

In order to deal with the questions raised in this study, three broad areas must be considered. First is the organizational structure, including the party, army, and other entities for the transmission of goals and policy. Second is the "human element," the primary group solidarity which controls and motivates the soldier. Third is the impact of the cadre leadership on the primary group and the resultant individual commitment toward engaging in combat as desired by the organization (Shils and Janowitz 1948; Etzioni 1961; George 1967).

The specific focus of research was where the primary group, the cadre-leader, and the organization came together in the PLA at the squad and platoon level. The PLA platoon and squad-level units were ideally suited for this approach because (1) the formal party and military organizations were evident at this level, (2) it was possible to see how primary group members responded as individuals within these organizations, and (3) the motivational techniques and the control mechanisms of the cadre were visible.

The Vietnamese PLA, like the Chinese PLA, did not allow small group ties to develop on an autonomous basis. Instead, the organizational structure of small groups was carefully controlled by means of the well-known 3-by-3 military cell system.[15] This organizational device was given two different emphases in the PLA: The three-man cell (*to tam tam*) acted as a formalized "buddy system" whose informal rules guided behavior on a face-to-face basis congruent with the requirements of the PLA (Kellen 1969); it served also as a surveillance mechanism which extended beyond the formal military and party organization into the very midst of the primary group itself. In this manner the three-man cell embodied the organizational principle of "responsibility." The soldier was never allowed to be an individual and was constantly reminded of his duties to his two buddies, to the unit, to the country, and to the party. By means of the three-man military cell adopted by General Giap, first in organizing the Vietminh and later the PLA in South Vietnam, the party succeeded in extending its organization to the roots of the PLA.

Studies indicate that the party was aware of the usefulness of additional organizational and psychological techniques which increased control over the primary group. Isolation of small units, secret surveillance, rotation of units, and leave policies are cited as examples of organizational devices used to induce behavior congruent with PLA policies (Kellen 1969, 1970; Donnell, Pauker, and Zasloff 1965). (These and other organizational characteristics within the PLA are discussed in Chapter 2.) Still other studies cite the tremendous impact of the *kiem thao* sessions (criticism of individual and unit activities) and the *phe binh* sessions (self-criticism) on members of the primary group. Still other psychological mechanisms such as voluntarism, public exhortation, and a sophisticated awards system provided the cadre with additional instruments of control.

Commitment to the party organization resulted from ideological indoctrination efforts. There appear to be no practical means of measuring the extent of indoctrination within the PLA, but indications are plentiful that indoctrination efforts had a significant effect. One study dealing with the "hard core," loyal Vietcong showed the impact of extensive indoctrination. Responses by the "hard core" interviewees were invariably the party line (Zasloff 1968), and suggest that the Vietnamese communists strived to create not just a "good soldier" but a "good communist soldier." This was achieved not by brief, superficial ideological training lectures but by the way of life that the soldier entered in the PLA. The thoroughly politicized cadre suffused the carefully structured small group with a pro-organization culture. To become a member of a small informal comradely group in the PLA meant acceptance of its organizational values, norms, ideology, and commitments, at least outwardly.

The degree to which the PLA was successful in its indoctrination effort is discussed in Chapter 5.

The structure and characteristics of the primary group within the PLA are discussed in Chapter 3. The relationship of the primary group to its larger secondary group, the Vietnamese people, and the impact of the secondary group on PLA motivation and control are covered in Chapter 4.

The techniques used by the cadre to motivate small group behavior toward the accomplishment of military objectives has been investigated by several studies. For example, the policy of isolation or unit compartmentalization together with a pervasive, cadre-implemented control and surveillance system were the main supports of unit cohesion in the PLA beginning with the escalation of 1965. These two factors are seen as having allowed the PLA to withstand tremendous setbacks and adverse developments (described in the next chapter) which might have caused a less cohesive army to disintegrate.

The PLA cadre are motivated by the prestige, recognition, and privileges which motivate any professional. In addition they are deeply affected by years of indoctrination. One interviewee assessed the effects of communist indoctrination on himself as follows:

> I lived in the resistance for eight years, and eight or nine years in the DRV, in a socialist world. It is not a political book which influenced me and formed my political ideas. I think that they grew in me from day to day. Each day a small quantity of socialist ideas entered me.[16]

The years it takes to develop such men are not regarded as excessive as they form the indispensable base for the "organizational weapon." According to Selznick, the necessary qualities of good communist cadres are absolute devotion and loyalty to the party and the closest possible contact with the masses. A party organization based on cadre "is a highly manipulatable skeleton organization of trained agents; it is sustained by political combat and is linked to the mass movement as its members become leaders of wider groups. . . ." (Selznick 1952).

The term *cadre* comes from the French where it means "frame." Cadre is also used in military terms to mean combat-hardened and experienced troops around whom an army can be developed. The PLA combined these two definitions to form what Douglas Pike calls the secret of Vietcong success, an organization of politicized cadre superbly suited to their military task. The role of the cadre in the PLA and the process by which it motivates and controls its soldiers is discussed in Chapter 5.

This study provides a description of the techniques and mechanisms relied upon within the PLA at the company, platoon, and squad level for the motivation and control of the Vietcong soldier in combat. The method utilized was significantly influenced by literature on why men fight which has emerged from studies of different armies in earlier wars.[17] This approach was further indicated by the many signs in the early and middle 1960s that the PLA was making its greatest efforts in the human areas of motivation and control. Thus, the questions asked and the sources used were picked to investigate thoroughly this aspect of the PLA.

Ideally, the most accurate way to investigate motivation and control in the PLA would be to observe PLA units in combat over an extended period of time. Because this was not possible the less desirable method of relying upon PLA defectors and POWs as "informants" on Vietcong morale and control systems was used. The source of this information was selected RAND Corporation interviews of PLA POWs and ralliers (defectors) between 1965 and 1967.[18] In all, the RAND Corporation interviewed approximately 2,400 Vietnamese between 1964 and 1968 who were familiar with PLA operations. These interviews cover an extremely wide number of subjects. Each subject or group of related subjects was the object of a specially designed questionnaire. The interviews used as the primary source for this study, some sixty-one from the "K Series," were specifically designed to investigate motivation and control in the PLA and were based on questionnaires designed for the purpose of studying motivation and control in a number of armies beginning as early as the 1940s.

As has been the case in the past, such an approach cannot be considered to draw upon a representative sample of the PLA or be representative of a particular unit of the PLA. Because of this limitation the "sample" of interviews utilized will not consider the interviewees as respondents whose answers may be analyzed using quantitative survey-research methods. Instead, each interviewee may be regarded only as a representative of the small unit he was assigned to within the PLA and, as such, may be considered to be a reliable and valid source of information concerning his motivation, the motivation of his comrades in the small, intimate, face-to-face group of which he was a member, and the group's relationship with the cadre.

There is, of course, a question of the reliability of any one interviewee. However, investigators who have read more than a thousand interviews concerning Vietcong motivation and morale were impressed with the consistency of the answers in a wide variety of areas (Kellen 1970). Furthermore, there was little reason to question most responses. Extensive efforts were made to ensure the interviewees that their answers would not affect their status as prisoners and that no harm would come to them for answering questions about their experiences in the PLA. Even those prisoners who defected from the PLA because of some overriding dissatisfaction often spoke admiringly and seemingly accurately about what they had observed in the PLA.

Therefore, when a number of interviews have similar responses or a similar attitude reoccurs several times in an area considered to be significant this was referred to by such qualifying terms as "all but one interviewee" or "most interviewees" stated that such and such was the case in their unit. The findings of this approach will in all probability not precisely fit any specific PLA unit. On the other hand, the advantage of this approach is that one does not have to rely on fragmented and incomplete data, which would be the case if data were collected by unit. These interviews are capable of providing the general dimensions of PLA motivation and control produced as a result of organizational, cadre, and primary group interaction.

In addition to the interviews, other primary sources of information have been used. Thousands of documents came into United States and Allied hands during the war. They were usually read at once for any immediate tactical value and the more significant were forwarded to Saigon for detailed analysis. From these, hundreds of documents pertaining to all aspects of Vietcong activities were translated into English by the U.S. Mission in Vietnam. These were generally titled *Viet Nam Documents and Research Notes*, and were available on request from the Counselor for Information, American Embassy, Saigon. Copies of the Vietnamese language originals were also available. From this broad source hundreds of captured documents were screened for this study. Those selected were generally of a routine nature, often intended for unrestricted distribution within the PLA and dealt with standards of behavior, roles of the party and cadre, rules, regulations, results of criticism meetings, and PLA organization, as well as captured letters and diaries. In addition, United States Army officers, including myself, have been eyewitnesses to much of what is described here.[19]

An extensive body of secondary works including books, monographs, and memoranda on Vietnam as well as literature on the motivation and control of armed forces has also been relied upon.

One final note concerns the time frame covered by the study. Because material from the sources described is sufficient for a thorough examination of motivation and control within the PLA from 1965 to 1967 this study is limited to this period.

NOTES

1. *Motivation* refers to human behavior caused by needs and values. *Control* is the accomplishment of a goal through the anticipation of predictable human behavior in circumstances which have been prearranged to some extent.

2. Morris Janowitz, "Hierarchy and Authority in the Military Establishment," in Amitai Etzioni, *A Sociological Reader* (New York: Holt, 1969), p. 211.

3. Ibid., p. 210.

4. As quoted in Alexander George, *The Chinese Communist Army in Action* (New York: Columbia University Press, 1967), p. 25.

5. The official name of the VC armed forces is the People's Liberation Armed Forces (PLAF). However, this study will use the terms popular with the VC soldiers. They referred to themselves as the People's Army (PA) or the People's Liberation Army (PLA).

6. Charles H. Cooley, *Social Organization* (New York: Charles Scribner's Sons, 1909), see Introduction.

7. Edward A. Shils and Morris Janowitz, "Cohesion and Disintegration in the Wehrmacht in World War II," *Public Opinion Quarterly* 12 (2), 1948, p. 281.

8. Ibid., p. 284.

9. Alexander L. George, "Primary Groups, Organization and Military Performance," *The Study of Leadership* (West Point, N.Y.: USMA Printing Plant, 1972), p. 19-3.

10. S. L. A. Marshall, *Men Against Fire* (New York: William Morrow, 1947), p. 42. Others also report on the phenomenon: Shils and Janowitz 1948; Little 1964.

11. Shils and Janowitz, "Cohesion and Disintegration," p. 284.

12. Morris Janowitz and R. Little, *Sociology and the Military Establishment* (New York: Russell Sage Foundation, 1965), p. 78.

13. Shils and Janowitz, "Cohesion and Disintegration," p. 287.

14. Dollard 1943; Homans 1946; S. L. A. Marshall 1947; Shils and Janowitz 1948; Stouffer 1949; Shils 1950; Little 1964; Janis 1971; George 1967. The same findings have also been shown to be true in the PLA (Kellen 1969, 1970; Elliot 1969).

15. The 3-by-3 military-cell system is a basic organizational principle for structuring PLA units. Each unit is one of three identical units which are all part of a higher unit which in turn is one of three identical units, etc. The 3-by-3 principle is discussed in Chapter 2.

16. Joseph J. Zasloff, *Political Motivation of the Viet Cong: The Political Regroupees,* RM-4703/2-ISA/ARPA (Santa Monica, Calif.: RAND Corporation, 1968), p. 40.

17. The seminal study of the Chinese People's Liberation Army by Alexander George has been important in shaping my thinking and influencing my method. George asked many of the same questions about the Chinese PLA that this study asks about the PLA. See George, *Chinese Communist Army.* There is also considerable evidence that the Vietnamese communists, especially General Giap, were strongly influenced by the Chinese in their construction of the PLA (Pike 1966; Tanham 1967; George 1967).

18. See Appendix A for fuller description of the interviews and information on the respondents.

19. Major William Brady, a prisoner of the PLA for six years, was an invaluable source in corroborating my findings and filling in detail on activities within PLA squad, platoon, and company-sized units. Major Jean Sauvageot, who served in Vietnam for seven years and speaks Vietnamese fluently, also contributed his knowledge. His personal experience in conducting countless interviews of PLA soldiers and his knowledge of PLA organization and control techniques also served to confirm the findings of this study.

1

The Setting

The North Vietnamese began preparations as early as January 1959 to use armed force in the South.[1] Citing several historical "betrayals" of the "Vietnamese People's Revolution," large numbers of PLA "regroupees" infiltrated from the North to reinforce a small group of active cadre in the South.[2] The immediate military objective was the neutralization of the strategic hamlet program, and defeat of the American-advised, -equipped, and -trained ARVN.[3] By the end of 1964, their efforts were so successful that an immediate commitment of American armed forces was necessary to prevent an imminent ARVN collapse. Thus began a massive American involvement in South Vietnam which, by the end of 1967, would see the commitment of 40 percent of all combat-ready U.S. divisions, one-half of U.S. tactical air power, and one-third of U.S. naval forces, costing in excess of $25 billion per year. Never before had such massive military power been continuously focused on such a limited area. In the face of such an overwhelming challenge, PLA reaction was outlined by its Chief of Staff, Van Tien Dung:

> . . .our arms and equipment were weaker than the enemy's. Thus we could only develop political and moral superiority, and only then have the courage to attack the enemy, only then dare to fight the enemy resolutely; only then could we stand solidly before all difficult trials created by the superior firepower that the enemy had brought into the war.[4]

In meeting the American challenge, the period from 1965 to 1967 was one of ever-increasing stress on the far-flung PLA organization.[5]

The year 1965 was a year of military crisis and momentous military decisions for the United States and the ARVN. The year began with a division-sized PLA attack already in progress against ARVN units at Binh Gia. It was viewed by General Giap as the first significant battle in the final phase of the war in the South. Increased infiltration of PLA units from North Vietnam into the highlands, along with other evidence, suggested that the PLA intended to cut South Vietnam in half from Pleiku in the west, to Qui Nhon on the eastern coast.

The South Vietnamese Army, its top leadership weakened by a series of coups and political upheavals and demoralized in the ranks, was judged by U.S. observers as being incapable of withstanding this PLA offensive. As an immediate response to the PLA threat, U.S. advisers received permission to employ U.S. air power, to include B-52s. While significant military results were achieved through the employment of air power, it failed to stop PLA progress. The ARVN continued to lose infantry battalions and district capitals at the rate of one per week. By October of 1965, the PLA had grouped three regiments in the Central Highlands adjacent to the Cambodian border. These were main force units which would provide the backbone for the major thrust to cut South Vietnam in half. In order to stop the PLA in the highlands, it was necessary to deploy the 1st Cavalry Division (air mobile) in the Central Highlands at An Khe, halfway between Pleiku and Qui Nhon.

The PLA began their offensive by attacking the Special Forces camp at Plei Me, southwest of Pleiku. In reaction, the 1st Cavalry Division was committed, leading to the battle of the Ia Drang Valley. Superior U.S. firepower and mobility led to a costly and bloody PLA defeat and succeeded in blunting the PLA drive to divide South Vietnam in half. This battle eased the military crisis for the ARVN, as U.S. units followed up, seeking out and engaging the PLA, leaving the ARVN to secure the population centers.

By the end of 1965, U.S. strength had risen to 148,000, which was the equivalent of five combat divisions plus accompanying support units. In addition, one South Korean division and a separate brigade arrived in South Vietnam.

Although set back by the deployment of the 1st Cavalry Division and other U.S. units, the PLA continued with determination. Its strength increased to 221,000 by the end of 1965, which included eight regiments infiltrated from North Vietnam during the year. During this period, PLA losses were 35,000 killed in action and an additional 6,000 captured.

While the year ended with an overall increase in PLA strength, the massive U.S. commitment succeeded in halting the increasing ARVN disintegration and blunting PLA initiative. In addition, the PLA was subjected, for the first time, to an unprecedented concentration of conventional firepower and military pressure which was to increase in intensity during the next two years.

The year 1966 saw the completion of the U.S. troop buildup and the beginning of major offensive operations. From the PLA side, the new battlefield realities prevented the pursuit of the "general warfare" phase and its strategy designed to split South Vietnam in half. While the PLA pondered its strategy, it was brought under intense pressure during the dry season by a series of Allied offensive operations emphasizing mobility and firepower. In an attempt to relieve the pressure, the PLA opened a new front just south of the DMZ in Quang Tri province. By March, the infiltration of the 324B and 341st PLA Divisions across the DMZ was well under way. In addition, other forces were infiltrated through Laos toward Hue. In all, the PLA committed or had in reserve in the DMZ area five or six divisions. Their purpose was to relieve the intense pressure from search-and-destroy operations elsewhere in South Vietnam. In the North, the PLA was favored by shorter lines of communication to North Vietnam, and by the presence of a sanctuary just across the DMZ. The Allied response was to rush sufficient forces north in order to contain that threat, while continuing to maintain intense pressure elsewhere in South Vietnam. As a result, 1966 saw numerous battles throughout South Vietnam. The largest, labeled Operation Attleboro by the Allies, was in Tay Ninh province. More than 22,000 U.S. and Allied troops were involved in the battle which lasted weeks and so completely defeated the PLA 9th Division that it was withdrawn from operations until the spring of 1967.

Still, 1966 was indecisive. Each side reinforced its strength in preparation for an even more intensive confrontation in 1967. By the end of 1966, U.S. forces had increased to 385,000 men. During the year, the PLA succeeded in infiltrating an additional 58,000 (equivalent to five divisions) for a total strength of approximately 370,000 men. This buildup was accompanied by intensified pressure on the PLA and resulted in an increase in defectors to the RVN's *Chieu Hoi* (defector) program. The total number of defectors for 1966, 20,242, was almost double the 11,124 for 1965.

The year 1967 developed into a year of unprecedented strain and pressure for the PLA. While 1965 was a year for blunting the PLA initiative, and 1966 a year for completing the U.S. troop and base buildup, 1967 saw the firepower and armed might of the U.S. brought to a sharp focus on the PLA. United States strategy was aimed at putting the PLA under such stress that it would cease to function as a military force. This strategy was outlined in early 1967 by General Westmoreland. "The only strategy which can defeat such an organization is one of unrelenting but discriminating military, political, and psychological pressure on his whole structure—at all levels. . . ."[6]

To accomplish this, the U.S. applied more conventional firepower in 1967 than had ever been previously concentrated in an area the size of South Vietnam. During 1967, U.S. strength rose to approximately 500,000 men. The number of Allied maneuver battalions rose to 278, with 28 tactical fighter squadrons in support. More than 3,000 helicopters gave Allied forces a previously unimagined mobility. Perhaps more significantly, the number of B-52 sorties increased sharply to 1,200.

With this firepower available, U.S. forces moved even farther from the population centers to challenge the PLA in areas which had long been under PLA control. With the exception of Operation Attleboro in 1966, these were the first such large-scale Allied offensive operations of the war. The PLA main force units were gradually forced into remote areas and jungle sanctuaries in order to maintain themselves as cohesive, organized units. Numerous PLA attacks and strategic moves were spoiled by massive B-52 raids while they were still in the assembly stage. It was a time of trial and challenge for the PLA as plans and strategies, adopted with high expectations, proved fruitless with enormous cost in personnel and supplies of all types.

In addition to the firepower listed, by the end of 1967 the Allied forces had massed a total force of 1,343,800 troops in South Vietnam. The combined effects of this military power increased PLA casualty rates to a high of five PLA casualties for each Allied casualty. This was from a low of 1.5 PLA casualties per Allied casualty at the beginning of 1965.[7] The casualty rate had a significant effect on the composition of the PLA. From a high of almost 100 percent at the beginning of 1965, the number of PLA battalions raised in the South decreased to just over 50 percent by the end of 1967. Moreover, PLA battalions originating in the South were given replacements from North Vietnam so that by the end of 1967, approximately 20 percent of their strength consisted of North Vietnamese soldiers.

By the end of 1967, not only the continued operation of the PLA, but its survival as well, seemed to be in question to many observers.[8] The sustained and ever-increasing military pressure between 1965 and 1967 exposed the PLA to external and internal forces experienced by few armies in the past, and survived by even fewer. This study examines the PLA organization, the individual soldier, relevant Vietnamese cultural characteristics, and the cadre-leader as critical factors necessary to an understanding of the remarkable performance and cohesion of the PLA during this period, and to its final victory in South Vietnam.

NOTES

1. According to the published resolutions of the Central Committee of the Viet-Nam Worker's Party meeting of December 1963, "The Resolution of the 15th Central Conference (Jan. 1959) and the subsequent Resolution of the Third Party National Congress have clearly set forth

the policy for the Revolution in the South and the (armed) struggle guideline. . . ." Document no. 96, *Viet Nam Documents and Research Notes*, "The Viet-Nam Worker's Party's 1963 Decision to Escalate the War in the South," American Embassy, Saigon, 1971, p. 3.

2. Expressing anger and disappointment at the failure to hold the elections for the reunification of Vietnam, large numbers of South Vietnamese, who were "regrouped" north after the victory of 1954, began to infiltrate south in 1959 to lead the armed struggle for "liberation and reunification." For a detailed and authoritative account of the "regroupees" and their significant role in "the Second Resistance," see J. J. Zasloff, *Political Motivation of the Viet Cong: The Political Regroupees*, RM-4703/2-ISA/ARPA (Santa Monica, Calif.: RAND Corporation, 1968), and *Origins of the Insurgency in South Vietnam, 1954-1960: The Role of the Southern Vietminh Cadres*, RM-5163/2-ISA/ARPA (Santa Monica, Calif.: RAND Corporation, 1968).

3. ARVN refers to Army of the Republic of Vietnam.

4. Van Tien Dung, "Ban ve nhung kinh nghiem xay-dung lucluong vutrang cach-mang cua Dang ta" (On Experiences in Building the Revolutionary Armed Strength of our Party). Taken from a paper presented at the American Political Science Association Convention in San Francisco, September 1975, "The Political Role and Development of the People's Army of Vietnam," by William S. Turley, Department of Political Science, Southern Illinois University, Carbondale, Ill.

5. Much of the review of United States and PLA strategy and action summarized in this chapter comes from personal recollection. The figures on relative troop strengths, sequence of events, etc., were taken from U. S. G. Sharp and W. C. Westmoreland, *Report on the War in Vietnam* (Washington, D.C.: U.S. Government Printing Office, 1968), pp. 107-207.

6. William C. Westmoreland, address before a Joint Session of Congress, April 28, 1967.

7. It is recognized that casualty figures reported by both sides were subject to question regarding the accuracy of the numbers stated. However, for the purposes of this study exact casualty figures are not necessary.

8. In addition to a number of Western observers, this opinion was held by many high-ranking officers from Warsaw Pact nations with whom I held numerous private discussions.

2 The Organization

Organization may be viewed as a means for accomplishing goals and objectives. Organization tends to minimize intragroup conflict and lessen the impact of individual behavior which deviates from the goals and objectives that the organization has established as worthwhile. By reducing uncertainty through prescribing acceptable forms of behavior and establishing expected roles for its members, the organization brings stability into human relationships and, particularly important for the PLA, made predictability of human actions possible in many situations.

Because of these attributes of organization, policy-makers can bring about desired goals and objectives. Selznick states that communists everywhere understand these principles and have consciously attempted to use them to their advantage. In *The Organizational Weapon*, he points out that the underlying assumption of communist movements "in all countries" is that the masses will move but "the direction in which they move will depend on the leadership." Control of leadership is the "decisive element" which all communist organizations seek. In this view, the party is a weapon to control mass movements. Its role is to link itself through "continuous political combat" to groups and movements by providing leadership for them.

In Vietnam, party control of the PLA was an extension of the "organizational weapon," much as Selnick described it.

The party goal was to gain control of Vietnam and for this purpose a military organization was indispensable. Therefore, the PLA organization will be discussed in terms of: party control of the PLA, especially at the company, platoon, and squad levels; the scalar structure from the highest decision-makers to the lowest-ranking soldiers; and the PLA's attempts to control totally all behavior within its ranks.

SCALAR ORGANIZATION

The People's Liberation Army includes three types of military organization within its ranks.[1] This is clear from the following statement by General Giap:

> Experience in carrying out the people's war over the past 25 years in our country has proved that the three categories of troops—(1) main force units, (2) regional forces, and (3) guerilla and self-defense militia forces—are the most appropriate organizational form to mobilize all people to fight the enemy.[2]

There was little organizational difference between main force units from North Vietnam and those originating in the South.[3] The internal organization and military practices of each were remarkably similar, and ethnic northern soldiers were assigned to main force units organized in the South when high casualties in southern units made this necessary.

The People's Army main force units were the primary combat force for the party, while local guerrilla forces functioned primarily as a village militia and regional units operated in an area generally no larger than a province.[4] The main force units more closely resembled the regular units of an army. They were mobile forces capable of engaging in modern warfare in a nation-wide contest.

Within the PLA, "party control over guns" was not an empty slogan. Countless PLA documents state that military objectives were subordinated to political goals. The party was represented at each organizational level and dominated the parallel military structure. At the top level, the Military Affairs Committee spoke with the authority of the overall command structure for South Vietnam, the Central Office for South Vietnam (COSVN).[5] In COSVN the military and party organizational structures found at all lower levels merged to form one combined political military authority (see Figure 1).

Below COSVN, main, provincial, and district forces, as well as local guerrilla units, were responsible to the appropriate level party organization (see Figure 2).

At each level a Military Affairs Committee (MAC) provided the critical link to the party through a party committee or chapter within the military units. The primary function of the MAC was to insure that tactical military operations conformed to party goals. The MACs followed the organizational principle of "substantive decentralization." At each level an MAC, aware of party goals and policy, adopted guidance from higher echelons to the local situation before issuing orders to the military headquarters at its particular level. This function is illustrated by the following passage from the captured notebook of cadreman Muoi Khan, which states that the MAC should:

> 1. Consider the resolutions from the Party Committee and and from the higher military echelon to determine the specific military tasks for the local area (region, province, district), in accordance with the general military requirements and those of the local movement. Determine specific missions and activities for the armed forces so that they will be performed in line with higher echelons and in accordance with the local situation.
>
> 2. On this basis, conduct military activities in the right direction and towards our objective. Attacks should meet the requirements of both the general and the local situations. . . .[6]

Figure 2 shows that the party committees within the military units were subject to two higher sources of party authority: the party committee at the next higher military echelon; and the corresponding level civilian party committee. Even though the possibility of conflicting orders was diminished because military operations were planned in great detail and rehearsed many

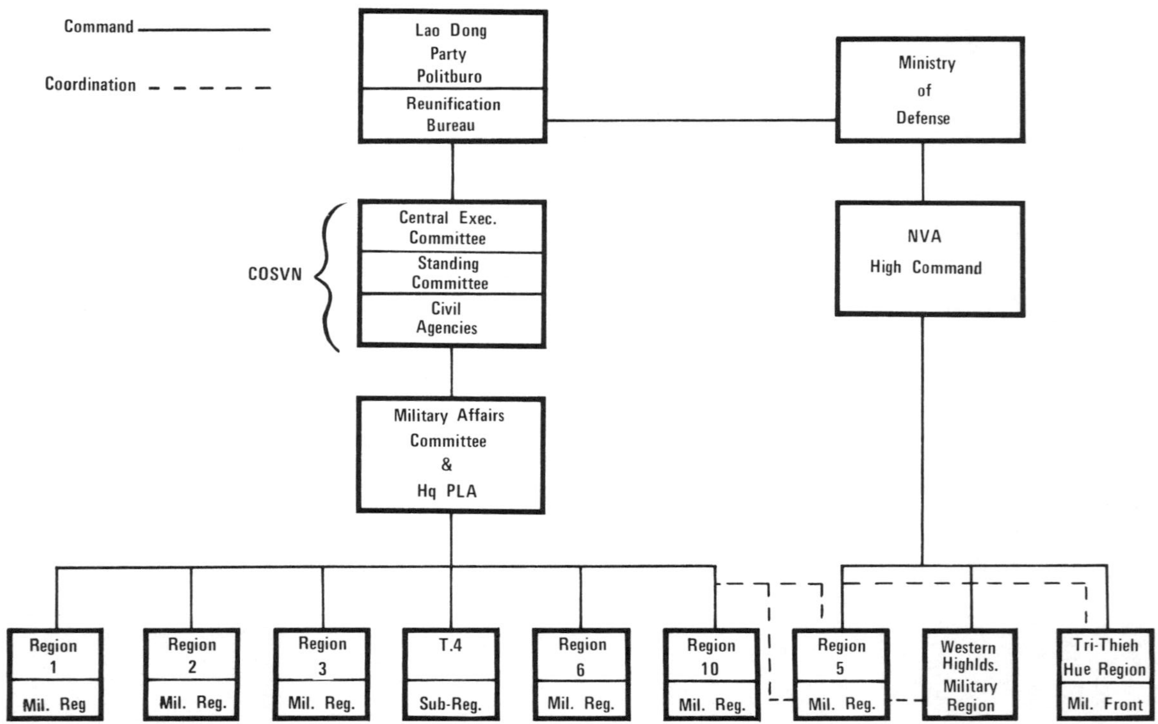

Figure 1. Party Political/Military Structure in South Vietnam, 1965-1967.

times, with time allowed to resolve differences, firm guidance was given. Under the heading "Relationships in Leadership," Muoi Khan's notes show:

> The Military Province (District, Village) Party Committee is subject to two command channels: one from the local Party Committee and the other from the Party Committee of the next higher military echelon. The local Party Committee directs the Party Committee for Military Affairs, using collective resolutions. . . . In case the orders and directives of the next higher military echelon are incompatible with the military responsibilities in the local areas, reports should be made to the Party Committee at the higher echelon, which will then make a decision. In an emergency, the military unit will carry out their task as suggested by the local Party Committee, then immediately report to the Party Committee of higher echelon and to the Party Committee of the next higher military echelon. The local Party Committee will be responsible for the decisions it makes.[7]

It is interesting to note that the "civilian" party organization was given precedence over the higher ranking military party organization. Strict guidance regulated the relationship between the party and military unit. Writing under the heading of the "relationship between the Party Committee and the unit head" (that is, the military commander), Muoi Khan noted:

> a. The Party Committee (i.e., political officer and other Party members) provides collective leadership, discusses and makes decisions on the following: matters relating to principles, policy and planning; matters concerning the execution of orders; matters concerning cadre affairs such as assignments, promotions, administration; matters concerning the arrangement and unification of the tasks assigned to the troops. . . .
>
> b. The political commissar and military commander are both heads of the unit. They share responsibility in all fields. The political commissar performs the role of Party secretary in the unit and takes charge of Party tasks. In Party and political tasks, he is subordinate to the next higher political commissar and political agency. In the military and administrative fields, he

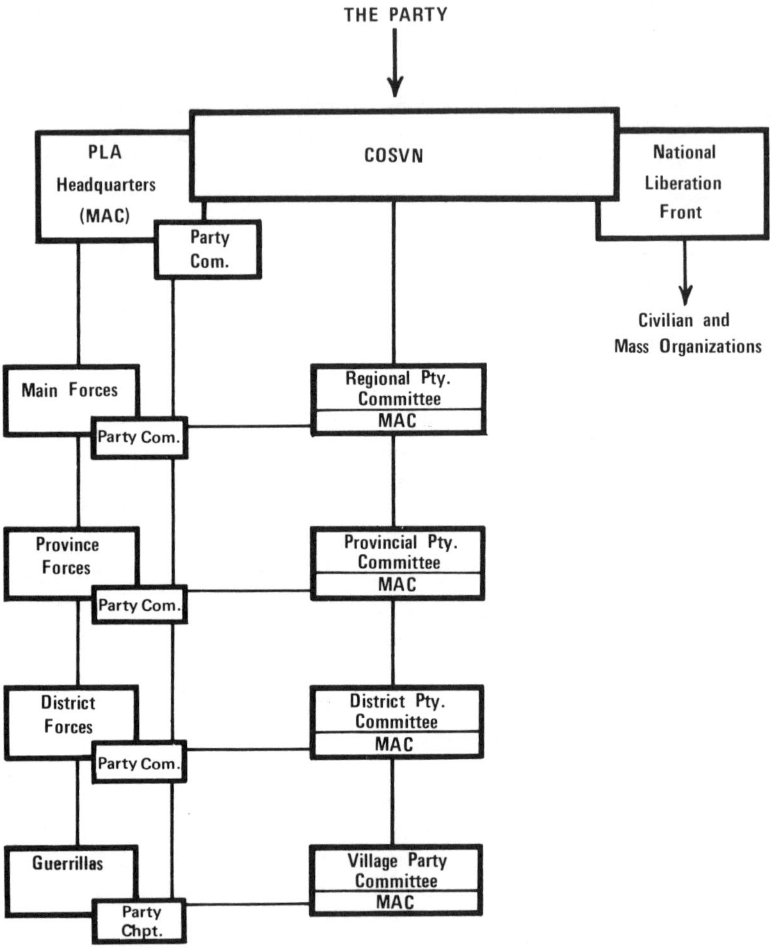

Figure 2. Party-PLA Relationship, 1965-1967.
MAC = Military Affairs Committee

> is subordinate to the military commander, the political commissar, and the military organ of the next higher echelon.
>
> The system of unit heads sharing responsibilities and of collective leadership by the Party Committee has been established by a long history of revolutionary struggle. It combines unified leadership with separation of responsibilities.[8]

Within the military headquarters emphasis was placed upon the accomplishment of military missions, but constraints on the military commander remained severe. The commander, always under the watchful eye of a party representative, was limited in his exercise of military initiative. The following response of an assistant company commander, replying to a question about the collective leadership at the company level, provides an illustration:

> The political officer was in charge of all aspects of the unit. Those were the fighters' living, the fighters' spirit, and the fighters' work. The Company Commander was in charge of security work (i.e., guards, etc.), postal service, and the troops' positions during operations. And, I, the Assistant Company Commander, was in charge of communication, transportation, and rear service.[9]

Military operations required approval of the appropriate party authority. Defensively, the commander issued orders only in an emergency situation. Even then these "contingencies" were usually well-planned and rehearsed in advance. The reliance upon collective party leadership and responsibility was so pervasive that military commanders were extremely reluctant to act beyond the letter of their orders. The following comments of a battalion officer, explaining the failure of his unit to execute a planned ambush, demonstrate this reluctance.

> Even though we had learned in advance that there would be 32 trucks carrying GVN reinforcements, we couldn't ambush them when they passed at 8 A.M. because the telephone wires connecting our position with the regimental command post, which was one kilometer behind us, were only then being

> installed; the regimental commander therefore couldn't give us orders in time as planned.[10]

These organizational relationships suggest a sharp compartmentalization of function and potential bureaucratic rigidity. However, as M.C. Conley points out, the party practice of assigning trusted individuals to key roles at several layers of authority prevented a rigid compartmentalizing.[11] This practice did not seem to interfere with separate organizational functions or negate the political distinction among separate organizational entities. As an example, an individual who was a provincial party secretary and a commanding general of the provincial military headquarters as well, was in an excellent position to coordinate the activities of both. However, he had difficulty in dominating their functional missions, since largely autonomous groups doing different jobs were involved in each organization.[12] In addition, such key individuals were always answerable to higher party authorities. Further, as Conley states, some "mutual distrust and competition" within the party and military organization were promoted since distrust served to root out deviance from party goals while competition served as a spur toward the accomplishment of these goals.[13]

This type of party organization is confirmed by other sources as well.[14] For example, captured official documents pertaining to the party and military organization of the 514th Battalion generally support this description of PLA organization. One document emphasizes the overriding importance of the party representative at each level of command. Another shows that the military command structure at the company level and below served primarily as a channel for technical advice. The following excerpt describes the function of the party cell within the demolition platoon of the 514th Battalion.

> The Party cell constitutes the organ that directly leads the platoon or squad in implementing the resolutions of the Party chapter. . . . Once a Party chapter resolution is put out, the unit commander in charge of military affairs and the unit commander in charge of political affairs should draw up a plan of action.[15]

No specific guidelines were presented by which the political officer, as a representative of the party, and the military commander, as the technician, were to reach agreement. The presumption was that the party cell was very knowledgeable in military matters and would make resolutions in accordance with the unit's capability. As this unit appeared to be highly trained and competent, the military commander served as a technical coordinator, executing the military course chosen by the party chapter. Another document states that in order for the party chapter to put forth timely and "relevant" resolutions, it must meet the following three conditions:

> 1. The Party chapter should have a very firm grasp of the political and ideological situation and of the performance of various tasks of the unit.
>
> 2. The Party chapter should have a very good knowledge of the military and political missions of the armed forces in general, and of the unit in particular.
>
> 3. The Party chapter should have a good understanding of. . . people's guerrilla warfare, and of the important policies of the party.[16]

Another excerpt demonstrates the major role of the party in planning a mission. The military commander did not have responsibility for formulating the final plan which was ultimately determined by political authority. Under the guidance of the party cell or chapter, "the entire unit meets to draw up a plan of action for the fighters in the unit to implement the resolution in an enlightened manner."[17] This procedure applied generally to all levels of organization within the PLA.

MAIN FORCE ORGANIZATION

Although most military units were controlled by a local party committee, main force units, which often ranged nationwide in their operations, were generally controlled by the Military Affairs Committee of COSVN.[18] However, care was taken that they adhere to local party policy wherever they operated. Main force units were cautioned that:

> When they operate in a given area, their relationship with the local Party committee is regulated according to the local situation. The main force units should accomplish the tasks assigned by higher echelons (COSVN) while maintaining close co-ordination with the local area throughout the time they are stationed in it.[19]

Main force units corresponded to World War II conventional Western forces of similar size and function. From the division through the regiment, battalion, company, platoon, and squad there was little difference in military organization. The internal organization of main force units followed the 3-by-3 principle adopted from the Chinese PLA by General Giap for the Vietminh and then later extended to the Vietnamese PLA. According to this principle, a manuever unit at any level is one of three similar units subordinate to a higher manuever unit which itself is one of three similar units (see Figure 3). (Some charts of VC organization are confusing on this point because they show a fourth, weapons support unit. However, this unit was usually employed by breaking it up and attaching its heavy weapons to the three manuever units.)

Although the 3-by-3 principle was used as a general guide in structuring units it was not invariably followed. Rather, the structure of an organization depended upon the availability of such elements as cadre, fighters, and equipment. This flexibility was especially important at higher levels (divisional and regimental) where deployment of large quantities of resources and manpower was required. Further, at lower levels, flexibility was also maintained where, for example, four fighters might be assigned to a three-man cell if there were insufficient suitable leaders.

In the PLA a party element provided the critical linkage to each military unit. At battalion level and above were party committees. At company level there was usually a party chapter (*chi bo*), at platoon level a party cell (*to dang*) was found, and at squad level the squad leader was usually a party member or candidate.[20] In some elite units with a large number of party members (such as the Demolition Platoon of the 514th Battalion) party chapters were often found at platoon level.

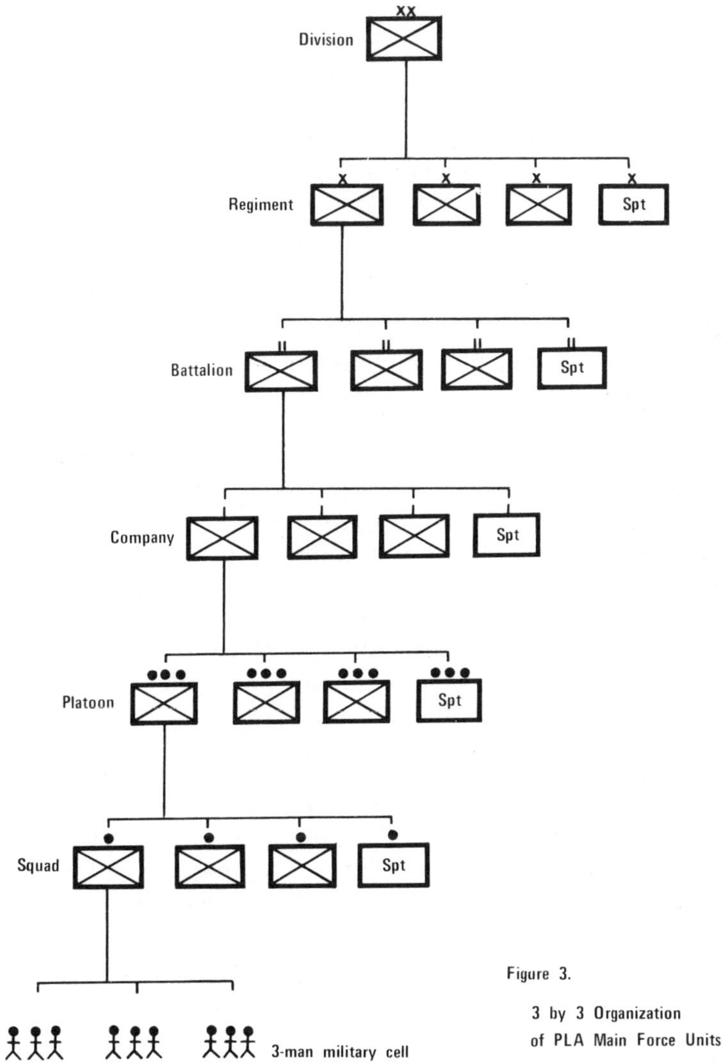

Figure 3. Three-by-three Organization of PLA Main Force Units.

Party committees or chapters were comprised of the party organization from the next lower military level. For example, a battalion party committee was made up of all company party chapters in the battalion. Company party chapters were, in turn, comprised of platoon party cells.

The dominant person in a committee or chapter was the party secretary (*bi thu*) who was assisted by a deputy secretary. As the number of persons in a company party chapter was often quite large, a chapter committee (*chi uy*) was formed which usually included the chapter secretary and his assistant plus the "leading" party members in the platoons and company. The battalion party committee was then formed from the "leading" company cadre along with most of the battalion command staff (*ban chi huy*). The party leader at the platoon cell level was the platoon political officer. The cell membership was made up of party members at the platoon and squad level.[21]

As the party organization at each level of command had the authority to go "over the head" of the military commander in all matters, it might appear that military initiative was stifled. Although some flexibility and desire to be innovative might have been lost this was usually not the case since the PLA military commander was usually a party member and he approached his tasks as a member of the party first and as a military "specialist" second.

The limits of direct party control were usually reached at the squad level in main force units. The squad leader was frequently a party member or a candidate. As its representative, the squad leader served as a vital link between the party organization and the ordinary soldier. The nonparty three-man cell (*to tam tam*) proved invaluable as an organizational aid to the squad leader for control and also for motivation. Unofficial "assistant squad leaders" were appointed from the most politically progressive soldiers to serve as leaders for these three-man cells. Often in elite main force units, the three-man cells were led by party candidates. The relationship of the squad leader as a party member to the three-man cell is made clear by the following excerpt:

> The Party chapter leads the Party members, and allocates to each of them the responsibility of keeping watch over the three-man cells and to report once every three days to the Party leaders about the situation and activities in the cell.[22]

Another major organization device used by the party to penetrate the ranks of the ordinary soldier was the party's Youth Group. Young men between the ages of sixteen and twenty-six were enrolled in an intensive indoctrination program designed to culminate in party membership. The Youth Group reinforced the three-man cell as an organizational control mechanism because its members became activists who provided surrogate leadership for the unit in the absence of party members.[23]

It is obvious that the party utilized considerable organization to control and subordinate all military personnel and activities toward its ends. At the vital point where the direct surveillance and control of party members ended at squad level, two primary control mechanisms, the three-man cell and the People's Revolutionary Party (PRP) Youth Group, were stressed as the primary means of extending party control into the ranks of ordinary soldiers.

NOTES

1. Because the origin of the PLA troops was an international political issue in the 1960s, four types of units were often discussed: (1) units of the North Vietnamese Army (NVA/PAVN); (2) southern main force units (*Quan Doi Chu Luc*); (3) territorial or regional units (*Po Doi Dia Phuong*); and (4) local guerrilla units (*Dan Quan Du-kich*). See, for example, Michael C. Conley, *The Communist Insurgent Infrastructure in South Vietnam* (Washington, D.C.: Center for Research in Social Structures, American University, 1966), p. 118.

2. Document no. 70, *Viet Nam Documents and Research Notes,* General Vo Nguyen Giap, "The Party's Military Line Is the Ever-Victorious Banner of People's War in Our Country," American Embassy, Saigon, 1969.

3. A significant role in achieving this organizational uniformity in VC military units was played by the regroupees who returned to the South with the mission of organizing and leading the PLA in SVN. For

a detailed account of the regroupees, see J. J. Zasloff, *Political Motivation of the Viet Cong: The Political Regroupees,* RM-4703/2-ISA/ARPA (Santa Monica, Calif.: RAND Corporation, May 1968).

4. In "The Party's Military Line Is the Ever-Victorious Banner of People's War in Our Country," General Giap states that the local guerrilla forces serve several functions for the party. Military defeat of the enemy did not appear to be the primary function. This was left to the regional forces and especially the main force units. Instead, the local guerrilla units emphasized: (1) developing and maintaining political bases; (2) production; and (3) providing regional and main force with fighters and cadre. Thus 70 percent of a local guerrilla's training was political and indoctrination. (See Conley, *Communist Infrastructure,* p. 118).

5. COSVN itself was subordinate to Hanoi, and reported directly to the Politburo of the Lao Dong party through the Reunification Bureau of the Politburo. There are indications that the role of the Reunification Bureau diminished as the Politburo increasingly dealt directly with COSVN on most issues. The Politburo maintained control over COSVN activities through a system of interlocking organizational directorships. The key leaders of COSVN were usually members of the Central Executive Committee or Politburo of the Lao Dong party. The Military Affairs Committee of COSVN (*Ban Quan Su*), during the period 1965-1967, acted as the headquarters of the People's Liberation Army and exercised command and control over main force, regional force, and guerrilla units. As a major suborganization of COSVN, the Military Affairs Committee (Military Headquarters) was headed by Tran Nam Trung, an alias given to each commander of the People's Liberation Army in the South. This position was always held by trusted northerners or regroupees high in the Lao Dong leadership. (For a detailed discussion of the organization, functions, and personalities of COSVN, see Document no. 40, *Viet Nam Documents.*

6. Document no. 34, *Viet Nam Documents.*

7. Document no. 34, *Viet Nam Documents,* p. 13.

8. Document no. 34, *Viet Nam Documents,* p. 9.

9. Interview K-11, RAND Corporation, Santa Monica, Calif., p. 15.

10. Interview AG-476.

11. Conley, *Communist Infrastructure,* p. 129.

12. Ibid., p. 130.

13. Ibid., p. 145.

14. This information was drawn primarily from captured documents and interviews of ralliers and prisoners of war. However, the discussion pertaining to the higher levels of the PLA organization is necessarily based upon a more limited number of sources. Some knowledge of internal COSVN organization, for example, is based only upon the comments of Colonel Tran Van Dac, a former political officer of COSVN's subregion 1 and a PLA veteran of 22 years' experience.

15. David and Mai Elliot, *Part Two: Party Organization—Documents of an Elite VC Delta Unit,* RM-5849-ISA/ARPA (Santa Monica, Calif.: RAND Corporation, 1969), p. 30.

16. Elliot, *Party Organization,* pp. 44-45.

17. Ibid., p. 45.

18. Except, according to Document no. 40, *Viet Nam Documents,* those main force units which operated in Military Region 5, the Western Highlands front (Pleiku, Kontum provinces), and the Tri-Thien-Hue front, which came directly under the command of Military Headquarters, Hanoi.

19. Document no. 40, *Viet Nam Documents,* p. 14.

20. The party chapter (*chi bo*) is equivalent to the Chinese PLA "Party Branch."

21. Elliot, *Party Organization,* p. 3.

22. Ibid., p. 85. It is important to note that the three-man military cell and the party cell were different entities. The military cell was usually limited to three members while the party cell usually consisted of more than three members and had different functions.

23. Ibid., pp. 7-8.

3

Small-Unit Structure

While primary group cohesion can work to the advantage of most armies, the PLA placed unusual emphasis on this approach. Through the three-man military cell (*to tam tam*) the PLA structured and controlled the primary group to a degree equaled in few armies.

PURPOSE OF THE THREE-MAN MILITARY CELL

Like the Chinese PLA before it, the Vietnamese PLA did not permit small group ties (that is, primary groups) to develop on an autonomous basis. Because apolitical group ties and loyalties were considered potentially subversive, the PLA worked intensively within the primary group to create "good communist soldiers." The three-man cell, indoctrinated with PLA party norms, was "molded" to embody the organizational principle of "responsibility." The PLA soldier was never permitted to be an individual; rather, he was constantly reminded of his responsibilities to his two comrades in the three-man cell, to the squad and platoon, to the people, and to the party. This is described in a captured COSVN document dealing with the "role," "mission," and "organization" of the three-man cell within the PLA.

> MISSION OF THE THREE-MAN-CELL
>
> The three-man-cell is responsible for improving its members so that they have a solid political and ideological

> background, good tactical and technical skills, and a good working ability. These members are grouped into a small firm "collective," so that in combat, they can always help one another dash forward to fight and gain achievements, regardless of the difficulties and dangers of the situation. . . . In combat and in the task of improving (our) companies, platoons, and squads, the three-man-cell plays a very important role. It helps form qualified individuals and strong units that can carry out any assigned mission. A qualified three-man-cell will be a good foundation to build strong squads, platoons, and companies.[1]

The comments of a former PLA corporal, corroborated by many other interviews, indicate the importance of the three-man cell to the individual soldier as well as the PLA.

Question: What do you think of the three-man-cell system? Did that system raise the morale and the fighting spirit of the VC troops?

Answer: *The three-man-cell system's purpose is to make people closely related and to help one another.* During fighting, everybody had the duty to take care of his wounded cell members or move his dead cell members out of the battle field. During a mission, people in a cell should stay close together and help one another in their joint duty. *The three-man-cell system helped the squad leader or platoon leader to have close control of his troops who operated separately.* For instance, during a mission in the plain, if a man stole something from the people, he would be criticized by his cell members (even) before they brought it up in a squad meeting. The cells usually met every evening to review the daily activities. . . [Emphasis mine].[2]

STRUCTURE OF THE THREE-MAN MILITARY CELL

The three-man military cell, usually composed only of fighters, was the basic building block of the 3-by-3 organizational principle. Squad leaders and above were not organized into military cells, although in the absence of a qualified cell leader, a squad leader might assume the functions of the military cell leader (that is, assistant squad leader) in addition to his normal duties.

The PLA strictly adhered to certain principles in organizing a three-man military cell. Party influence had to be strong. Considerable attention was paid to the selection, training, and guidance of assistant squad leaders, as they were the direct representatives of the organization. These points are revealed in the following portion of a COSVN document.

ORGANIZATION AND ACTIVITY OF THE THREE-MAN-CELL:

a. *Organization:*. . .We must organize the three-man-cell based on the following principles:

. . .we should pick three men who are most closely related to one another in their functions, and form them into a cell. (In special cases, we can organize cells of two or four men.)

We should organize them in such a way that the old members can help the new ones, and the qualified members can help the unqualified ones. We should insert Party and Group members into every cell to insure its quality.

. . .The assistant squad leader should be the cell leader, while the squad leader himself should be free to exercise overall command.

. . .the designation of cell leaders is based on the recommendations of squad and platoon cadre. . . .

b. *Scope of activities:* include discussions on methods to carry out the combat tasks and missions assigned to each platoon, squad, cell, and members; and examination of the physical and mental life of each individual; and his personal feelings and desires. . . .

The appropriate time for such discussions is during off-duty hours of a workday. However, we should be very flexible (concerning this matter of time) when the unit is on continuous combat duties. These discussions can be conducted every day, during off-duty hours in the evening, during the 10-minute breaks of a training session, or at the last halt of a movement near the bivouacking area. We can even conduct them while our units are conducting a combat operation or about to withdraw, as the situation permits, in order to strengthen our troops' morale. The themes of discussion must be short, simple, realistic, and in keeping with the situation and mission of the cell. . . .

c. *Cell leader training:*. . .The improvement of the three-man cell leader's capability is an even more important task. He must be thoroughly indoctrinated, both in politics and ideology, to become a nucleus to rally the men in his cell around him. He must always set a good example either when carrying out internal activities, conducting training, or performing missions; be courageous and resourceful in combat and capable of providing effective leadership for his cell under critical circumstance.

The three-man cell leader should be trained in a number of methods of controlling the ideological attitude of the men in his cell, including the method of approaching his members and finding out their thoughts through exchange of personal feelings.

He should know how to lead new soldiers assigned to his cell to enable them to quickly understand the situation, mission, and traditions of his cell and unit.

He should know how to conduct cell internal activities, training, and combat. . . .We should select a good cell leader to use as a model for other cell leaders to learn from and to exchange experiences with.

In addition to this training task, our cadre must watch each individual and cell closely, and carefully to find out their strengths and weaknesses in order to provide prompt motivation and correction. . . .[3]

FUNCTIONS OF THE THREE-MAN MILITARY CELL

The three-man military cell served four basic functions: (1) as a "buddy group" capable of satisfying the basic needs of the PLA soldier (see Introduction); (2) as a primary group within which the party/PLA representative insured that group norms were congruent with party goals (see Introduction and Chapter 5); (3) as the basic mechanism within the extensive PLA surveillance and reporting system (see Chapter 5); and (4) as an "autonomous," yet disciplined, fire and maneuver unit in combat, especially suited for guerrilla warfare.

The following COSVN document describes these functions.

> THE POSITION AND ROLE OF THE THREE-MAN-CELL
>
> . . .the three-man-cell is characterized by mutual aid between members which is based on. . .mutual affection among comrades. . .they take care of one another's moral and physical lives. . .all members help one another to heighten their political and ideological awareness, and their fighting spirit, promote their working zeal, and maintain the good qualities, ethics, and behavior of revolutionary cadre and soldiers. . . . The three-man-cell is suitable. . .for approaching members and finding out their thoughts through exchange of personal feelings. . .and for applying all the tactics of our army. Every combat and tactical initiative can be cleverly and skillfully applied through the three-man-cell.[4]

Specific guidance was given to the cadre on the desired behavior for all members of the military cell in performing these functions: They were told that all members should:

> . . .promote a high spirit of solidarity, based on the love of the (working) class, and esprit-de-corps. *They should treat one another as brothers.* They must thoroughly understand the family situations, personal feelings, ideologies, desires, and the strong and weak points of one another. They must love, help, and trust one another, be open to one another, and help one another strengthen (their) ideological stand.
>
> *They must positively and voluntarily help one another in their daily activities*, such as making their beds, sewing, washing their clothes, and preparing their food when they are sick or on a mission in the field. *They should help one another to follow the behavior and ethics of a revolutionary soldier, and strictly observe military discipline and follow the Party's policy.* In the political and cultural indoctrination, they should help one another to thoroughly understand the lesson, make proper comments, and exchange ideas. In military training, they should help one another to master every subject, (clearly assess) the strong and weak points of each cell member, and understand the purpose and requirement (of training). During movements and when bivouacking, they should help one another to carry and maintain weapons,

> equipment, funds, and other materials of the cell, and construct fortifications. *In combat, they should motivate one another to completely submit themselves to the battlefield discipline, and be courageous, resourceful and united to effectively coordinate with and support one another to destroy the enemy.* They should strive to help one another should any of them be wounded or killed in action [emphasis mine].[5]

Most interviews confirm the role played by the three-man military cell. Buddy relations are revealed by an excerpt from the following interview:

> The purpose of the three-man cell is to give constant assistance to each other. *In training, the good help the bad. In combat, two experienced men follow and help the new recruit. In normal life, the three men share the joys and sorrows.* The three-man cell exists from the war against the French. *It usually unites three best friends* [emphasis mine].[6]

Cadre direction of the military cell in combat was indicated by the following answer of a PLA private:

Question: Do you think the three-man cell system helps the men's fighting spirit?

Answer: Sincerely, it helped the men both in combat and in daily work. *For instance, in combat the cell-leader directly commanded the two other members of the cell while the (other) commanders were only indirectly in command.* By telling the two other men to carry out the attack, one on his right and the other on his left, the cell-leader was able to coordinate their actions and keep an eye on his cell [emphasis mine].[7]

The same private explained the surveillance and reporting function of the three-man military cell:

> When a three-man cell went on a mission in rural areas, the cell leader watched over the two other members of the cell, and the latter also kept an eye on the cell leader. If the cell leader did something wrong, the two members of his cell would report

> it to higher cadres to have him demoted. If the two other members were ill-behaved, the cell leader also reported it to higher cadres. This is their (the three-man cell) collective command and individual accountability policy.[8]

Primary group interaction was remarkably "democratic" within the prescribed party/PLA guidelines. As long as members remained within these guidelines their individual behavior suggested a "participative" attitude, in Gabriel Almond's terms, rather than a "subject" attitude, as one might suspect.[9] For example, one private stated, "if the leaders wanted to move the company to a certain place, they asked the fighters about the place. If the majority didn't agree to go there, the leaders would agree with the majority."[10] Despite this "fighter's" impression that the rank and file had a veto on unit activities, this was not always the case. Usually, a detailed and lengthy discussion of the forthcoming operation under the guidance and influence of the cadre resulted in a commitment of the fighters to the plan. This process of "democratic centralism" contributed to the success of planned operations. Through the psychological mechanism of cognitive dissonance, the fighters came to view the plan as their plan.[11] A cadre explained the system as follows:

Question: Were the fighters given a chance to discuss and criticize a plan of operation before the operation?

Answer: . . .yes, they were given the chance to discuss and criticize. The idea was to get unity of command and action during the operation. Before any operation, a few among us would be sent out to make a study and survey of the battlefield, and then a plan of operation would be drawn up and presented to all the men in the unit. *Each would then be given a chance to contribute ideas and suggestions.* Each squad, each man, would be assigned one task, while the whole unit would be told what action to take if the enemy was to take such-and-such a position. In other words, all the possibilities were considered. During such a meeting, *it was the cadres' duty to*

> *inform their men about the possible development of the action, but it was also the fighters' duty to contribute to the plan by advancing suggestions or criticizing what had been put forward. Thus, the final decision concerning an operation or attack was very often the result of a collective discussion in which each member had contributed his opinion or suggestion.* In certain cases, however, such as emergencies. . . it was the leaders of the unit who, alone, decided upon the operation. . . . [emphasis mine].[12]

Thus the three-man military cell (that is, the primary group) played a central role in the organization and functioning of the PLA. However, certain cultural characteristics of the Vietnamese people also contributed to primary group cohesion.

NOTES

1. Document no. 102, *Viet Nam Documents and Research Notes,* COSVN document on "Guidance on Organizing, Consolidating and Conducting Activities of the Three-Man Cell," American Embassy, Saigon, 1972, p. 44.
2. Interview K-27, RAND Corporation, Santa Monica, Calif., p. 18.
3. Document no. 102, *Viet Nam Documents,* pp. 45-47.
4. Ibid., pp. 43-44.
5. Ibid., pp. 44-45.
6. Interview K-9, p. 18.
7. Interview K-16, p. 22.
8. Ibid., p. 23. (The PLA system of surveillance and reporting and the role of the military cell in this system is discussed in Chapter 5.
9. Gabriel A. Almond and Sidney Verba, *The Civic Culture* (Boston: Little, Brown, 1965).
10. Interview K-19, p. 15.
11. Paul F. Secord and Carl W. Backman, "Festinger's Theory of Cognitive Dissonance," in *The Study of Leadership* (West Point, N.Y.: USMA Printing Plant, 1971), pp. 25-36.
12. Interview K-5, pp. 32-33.

4

Cultural Factors Affecting PLA Cohesion

The concept of the secondary group (*Gesellschaft*) also influenced PLA cohesion. *Gesellschaft* may be defined as impersonal relationships within a large organized group.[1] A secondary group is too large to function on the intimate face-to-face basis of the primary group, yet it also has cultural norms and values which guide the interactions and decisions of its members. These cultural values arise through time and may be traced to such factors within the larger group as race, language, history, and religion. These factors allow communication and interaction over many common areas necessary for the efficient conduct of group and individual business.

It has been proposed that PLA fighting units operated effectively because the party bound the individual soldier to the expectations and demands of the group by means of primary group ability to meet basic individual needs of security and esteem. Primary group cohesion in turn is greatly facilitated by the degree to which its members share common cultural values and characteristics.[2] In other words, if soldiers in a small unit are from a relatively homogeneous secondary group, cohesion among its members is likely to be enhanced. On the other hand, heterogeneous characteristics in race, religion, history, and language tend to hinder cohesion.[3]

A common culture among members of a secondary group provides, according to Shils, "one of the most important preconditions for the formation of primary groups which have. . .

a positive and immediate function in strengthening the soldier's will to exert himself under dangerous conditions."[4]

Therefore, the party manipulated or accommodated several salient Vietnamese secondary group characteristics to create an internalized discipline within the PLA. In doing so, the PLA had to accommodate or neutralize those effects of Vietnamese culture which tended to militate against discipline.

POTENTIAL NATIONALISM IN VIETNAM

Nationalism may be defined as:

> A belief on the part of a large group of people that they constitute a community, called a nation, that is entitled to independent statehood and the willingness of that people to grant their nation a primary and the primary terminal loyalty.[5]

The party vigorously sought to exploit Vietnamese nationalism within the primary groups of the PLA. The broad dimensions of potential nationalism available to the PLA were significant. This potential is made clearer by the following table which arbitrarily assigns values to several criteria for assessing nationalism.[6]

Table 1 Potential for Vietnamese Nationalism

1. Large Enough Group	++
2. Sufficient Territory	++
3. Common and Unique History	++
4. Common and Unique Culture	++
5. Common and Unique Language	++
6. Common and Unique Religion	—
7. Common and Unique Race	+
8. Nation is Primary Loyalty for Elite	+
9. % Population Aware of Nation	++

Code: ++ indicates very strong potential for nationalism
+ indicates strong potential for nationalism
— indicates weak potential for nationalism
—— indicates very weak potential for nationalism

Using the preceding definition of nationalism and the accompanying criteria the potential available to the PLA for mobilizing the forces of nationalism within its ranks can be examined. Two prime factors are *the group's numbers and its territory size.* With a population of at least 34 million in 1962—18 million in the North and 16 million in the South—the Vietnamese certainly are numerous enough and their territory of 127,000 square miles is sufficient to form the basis of a nation. A country of geographic diversity, Vietnam is 1,400 miles long, 39 miles wide (at the 17th parallel), with two distinctive climates and highlands and lowlands. It has been described as two rice bowls (the Red River and Mekong deltas) connected by an S-shaped bamboo pole (the thin coastal plain). Vietnamese often refer to their country as a chain whose strength is dependent upon the narrow links nestled between the mountains and sea which connect the population centers of the North and South.

Another factor contributing to nationalism is *a sense of common and unique history and shared values.* Generally, a people's history will be a source of common values and a sense of common past will be a cohesive force, especially if it includes periods of trial. Even more essential is an expectation of a common history in the future. The Vietnamese have both. Dating from 208 B.C., Vietnam has had a history of constant struggle against foreign domination, internal rebellion, and expansion to the south into the lands of the Champa and Khymer.[7]

The Vietnamese people's knowledge of their history has been strengthened by reliance on legend.[8] The telling and retelling of tales by poets, grandparents, and parents perpetuates Vietnamese history. "Vietnam: it is all the legends narrated on winter evenings, in the glimmer of an oil lamp. . .a country of poetry and of wisdom. . . ."[9] While historical accuracy may suffer in the telling, their popularity as a favored pastime has contributed immensely to the formation of the Vietnamese national identity.[10] Such a verbal "history," passed from generation to generation by the family, also imparts values and potential guidance. For example, Tran Hung Dao instructed, as he lay dying and the emperor came to ask his advice in case the enemy from the north (China) invaded again: "The army must have one soul like the father and son in the family. It is vital to treat the people with humanity, to achieve deep roots and a lasting base."[11]

Vietnamese literature and history reached even the lowest peasant by word of mouth. Ellen Hammer describes how minstrels carry in song the history of the nation and the exploits of its favorite heroes. In Hue: ". . .during the long nights of the hot summer, boatmen on the River of Perfumes still sing nostalgic songs about the emperors Ham Nghi and Duy Tam and how they resisted the French during the colonial period and were defeated."[12] Companies of actors traveled through the countryside performing plays depicting episodes from Vietnamese history. This technique, used by the party to spread its interpretation of Vietnamese history, was greatly enjoyed by the PLA soldier.[13]

The communist effort to control the interpretation of Vietnamese history has been extensive. While the communists are content simply to claim Vietnamese history prior to 1930 as part of their heritage, they have made considerable effort to influence the subsequent record.[14] History lessons, broadcast to the PLA in the South over Radio Hanoi, emphasized "historical" events designed to elicit a nationalistic response.

A common and unique language also promotes nationalism. During their 1,000-year domination, the Vietnamese assimilated much from the civilization of their Chinese overlords. Two Vietnamese cultural survivals from this period were the legacy of independence (*doc lap*) and the Vietnamese written language (*chu nom*), a system of transcription based upon Chinese characters.[15] In the nineteenth century, a new and even more distinctly Vietnamese writing system (*quoc ngy*), which relied upon a romanized translation of spoken Vietnamese, was adopted throughout the country. The leaders behind this new transcription were Vietnamese intellectuals, exposed to the West, who realized that the road to modernization, the secrets of technology, and independence lay in taking what was useful from the West. It was in the Vietnamese spirit of *doc lap*, or independence, that *quoc ngy* was introduced into Vietnam.

A sense of belonging to *a unique ethnic group or race is also a strong promoter of nationalism.* Here again the Vietnamese rank high. They trace their origins to 500 B.C. when several clans living in the Yangtze River region of China came under strong pressure from the Chinese to assimilate. The Nam (South) Viet clans migrated southward rather than submit to assimilation as did clans which remained. The Nam Viets, claiming to be ethnically pure, moved into the Red River Delta from where they have expanded to the South.[16] Not traditionally included in their group by the Vietnamese are several ethnic minorities which make up approximately 15 percent of the population of Vietnam.[17] These consist of highland aborigines, overseas Chinese, Chams, and Khmers.[18]

Although Vietnamese culture is characterized by great uniformity when compared with surrounding cultures, there are some areas of diversity. Among the most significant are the four great philosophies and religions imported from abroad: Confucianism, Taoism, Buddhism, and Christianity. The first three, having interacted for hundreds of years, have produced unique blends.[19] Christianity, primarily Catholicism, has had an important impact on a small but influential minority. The Cao Dai and the Hoa Hao religious sects increased the cultural diversity in South Vietnam. While the values of these religions were generally compatible with Vietnamese nationalism, interreligious conflict did occur. Thus, religion and parochial values do not strongly enhance the appeal of nationalism in Vietnam.

The great length and narrowness of the country, with its inherent difficulties of communication and its diverse climates, have created regional differences.[20] Pike claims southerners are seen by other Vietnamese as "lazy, slow-witted and boorish," the Central Vietnamese as "overly traditional, hidebound, and too political," and the northerners as "aggressive, warlike and money hungry."[21]

An especially significant factor contributing to potential Vietnamese nationalism is the *primary loyalty among the Vietnamese elite to the nation.* Within the PLA the "elite," of course, are the cadre. The question of whether party leaders were nationalists first or communists first is not important. Although evidence has been mustered to support both views, it is safe to assert that nationalism and communism were intertwined in Vietnam and not contradictory. Within the PLA, lower-ranking soldiers perceived their immediate leaders to be nationalists rather than communists for two reasons: (1) the ordinary PLA soldier was not personally aware of midrange or even top-level party leaders other than Ho Chi Minh himself; (2) lower cadre explanations of why it was necessary to fight the government of Vietnam (GVN) and the Americans were usually couched in terms of Vietnamese nationalism.

Evidence of this nationalist appeal is available from a variety of sources. Conversations with an American officer, who was a POW of the PLA, confirm the anti-American Vietnamese nationalist appeal in the various PLA units he observed over the six years of his captivity.[22] He became aware of this almost immediately after his capture. During the period prior to his capture he had formed close relationships with several members of the ARVN unit he was advising. Soon after his capture he and an ARVN lieutenant plus several ARVN privates formed a natural "we-they" group against their PLA captors. However, within weeks his PLA captors, using an anti-American appeal, had created a new "we" group, and the ARVN privates, after a six-month period of "retraining," were "allowed" to join the PLA. The American officer was isolated and even attacked by his former ARVN friends who accused him of "crimes against the Vietnamese people."

Within PLA units the message was similar:

> . . .the Americans were like the French before. The French came here because, according to them, the Vietnamese people were stupid and needed their help. Actually they came here to rule over the Vietnamese people. Now the Americans aren't much different from the French. . . .[23]

All PLA troops knew of Ho Chi Minh, referring to him as "uncle" or Ho Chi Minh, through his efforts for Vietnamese independence against the French and later against the Americans, rather than for his position in the party. Ho's reputation as a nationalist developed over the years. He spoke to the people in nationalist, not class or party, terms. He is often quoted as stating, "My party is my country. . .my program is independence." His slogans were "independence or death"; " Vietnam to the Vietnamese"; "Death rather than return to slavery."[24]

The final factor strengthening Vietnamese nationalism was *the Vietnamese people's perception of Vietnam as a unique, distinctive, and viable nation.* Vietnamese children learn national legends from parents, and in schools they learn of the Vietnamese nation and its uniqueness. For example, Tran Hung Dao's proclamation to his troops on the eve of battle has become a classic of Vietnamese literature which is studied in schools.[25] It celebrates the country's resistance to the Mongol invaders and speaks proudly of Vietnamese independence. When asked for the major factor in his defeat of the Mongols, Tran Hung Dao stated that an ". . .army must have one soul like the father and son in the family. It is vital to treat the people with humanity, to achieve deep roots and a lasting base."[26] This is a phrase frequently repeated by the PLA today.

Possibly the most popular legend in Vietnam concerns Le Loi, a national hero who freed Vietnam from Chinese rule. As part of his strategy Le Loi made famous a statement by the well-known Vietnamese poet, Nguyen Trai, which each Vietnamese child learns:

> Our people long ago established Vietnam as an independent nation with its own civilization. We have our own civilization. We have our own mountains and our own rivers, our own customs and traditions, and these are different from those of the foreign country to the north [China].[27]

VIETNAMESE GROUP ORIENTATION

Another Vietnamese characteristic that contributed to discipline in the PLA is the group orientation of Vietnamese society. Unlike Western cultures where individualism is often extolled and where behavior contrary to group norms often gains respect, individual Vietnamese behavior is more often the result of group demands and expectations. The Vietnamese live in a society dominated by primary groups. As Paul Mus notes, the Vietnamese ". . .have abundant group spirit, not only in their family life, but also in their village communities [which are usually composed of kin]."[28]

The extended family or the kinship system is the bedrock of the Vietnamese social system. The Vietnamese learn that loyalty to the group, especially to the family, is the primary virtue.[29] This is a manifestation of the Confucian ethic of *hieu* or filial piety. Centuries of *hieu* indoctrination have, according to Pike, made the Vietnamese "organizational animals," almost completely dependent upon the group to which they give their loyalty in return for the security and the satisfaction of other basic needs it offers.[30]

In their organization and techniques of control, the PLA understood that a soldier's strong attachment to his family and village made it difficult to organize main force units capable of operating anywhere in South Vietnam. The ingenious PLA technique for dealing with this problem demonstrated its success in harnessing the power of the group over the individual. Aware of the difficulties of drafting a man directly into main force units, the PLA first inducted him into a local guerrilla force, assigning him to his home area and allowing him to live with his family when possible. Next, the soldier was encouraged to develop loyalty toward a new group. At this point the soldier, along with his group, was ready for transfer to a main force unit.[31] A PLA indoctrination booklet described main force units as being formed of "young men with combat experience in the Guerrilla Popular Army or territorial forces who have been indoctrinated in ideas of the revolution and class struggle, in organizational methods, and in combat techniques. . . ."[32]

In the main force unit the PLA soldier depended upon his new three-man cell which served as a surrogate kinship group that satisfied his strong need to belong. Since main force units usually operated far from the soldiers' home villages, their security and social needs previously provided by the family and village kinship system were met by the three-man military cell.

THE HIERARCHICAL DOMINANCE OF VIETNAMESE SOCIETY

A third cultural characteristic which reinforced PLA control was the tremendous group pressure to conform to group expectations and to avoid controversy and conflict within the group. This "subject" orientation toward the group once again stems from the dominance of the group, especially the family, in Vietnamese society.[33] Vietnamese are raised in a hierarchical group in which:

> The behavior of the individuals in the family and with other families is bound by all sorts of rules and precedents. . . .The authority of the grandparents, the father, the mother, and eldest son is absolute, conditioned only by deference to the spirits of their ancestors. The rest of the family must take orders and carry them out. . .*and it is still the duty of members of the family to make the greatest sacrifices if it is necessary to save the honor. The individual is of minor importance.* The family is the only reality, and the father the only ruler [emphasis mine].[34]

Just as the PLA provided a surrogate kinship group for the soldier it also provided a group hierarchy, headed by a cadreman who, in effect, became a surrogate father and as such achieved a powerful position of influence.[35] As in the kinship group, the PLA hierarchy emphasized conformity and avoidance of conflict, under the authority of the cadre.

THE IMPORTANCE OF FACE

Another significant Vietnamese cultural characteristic is the concept of "face." Once again this cultural trait is based on the tremendous power of the group over the individual. The specific nature of face is illustrated by the following comments of military officers from Vietnam, Cambodia, and Thailand during a discussion on the meaning of face and loss-of-face.[36] One officer said it was the humiliation felt when an officer failed in a mission. Another related it to maintaining a reputation. Still another saw it as the violation of a taboo which would bring disapproval and shame to the individual. Implicit in each of these comments is a reference to group sanctions if expectations were not met.

Discussing the intense nature of face in Vietnam and the severity of sanctions for loss-of-face, McAlister has commented upon ". . .the number of cases in which loss-of-face resulted in murder, and even more, in suicide. This violence occurred because 'face' is not just another word for vanity or social pretension; it is the very character of one's being."[37]

The PLA uses the Vietnamese concern with face to control its soldiers. This is done primarily through self-criticism (*kiem thao*) sessions and, in some units, requiring public vows before the group.[38] A PLA cadre, in response to a question concerning the effectiveness of *kiem thao*, said "In self-criticism sessions, I promised I would make corrections, and after that I felt relaxed. Criticism is a good method, and it is a token of mutual affection in a unit."[39] Thus, the soldier promised to correct his behavior, implying future adherence to group norms (conformity) in exchange for the implicit promise of group acceptance (security).

Throughout the interviews, reference is made by PLA POWs to "correct" and "incorrect" thoughts. A PLA private stated:

> When a person whose defects were not criticized made a mistake or had a shortcoming, he could never realize his mistake and thus could never correct his defects. So I think the criticism is good. To be criticized means to receive assistance from the men in the squad. *Without criticism one might not know one's shortcomings* and consequently make even bigger mistakes [emphasis mine].[40]

Again, this response makes clear that all behavior was subject to evaluation by the unit as "correct" or "incorrect."

THE VIETNAMESE DRIVE FOR STATUS

Another characteristic used by the PLA to control behavior is the Vietnamese drive for status. Although not discernable in all soldiers, the interest in status was evident in many POW observations. The PLA offered the individual a career which would meet his need for advancement and self-esteem, in short, for status. The PLA was largely a meritocracy and advancement was, in large part, determined by demonstrated performance in meeting PLA and party goals.[41] The Vietnamese desire for status is apparent in the case of a POW who stated he "rallied" to the GVN side because his advancement in the PLA was limited. He stated:

> I was dissatisfied because I wasn't admitted to the Party. Because I wasn't a Party member, I couldn't be promoted. I noticed I was more capable than a few other men, but I wasn't promoted. . .(respondent paused). . .In the army they always know about the background of each man. I couldn't be promoted because they maintained that I was a bourgeois element.[42]

VIETNAMESE FATALISM

In controlling its soldiers the PLA made effective use of another characteristic of Vietnamese culture, fatalism. In describing the spiritual forces of Vietnam a well-known Vietnamese poet and essayist, Tran Van Tung, states that the people, dominated by centuries of struggle to conquer and live in Vietnam, have practiced the stoic and ascetic teachings of Buddhism which led the Vietnamese to believe that: "since it is impossible to do away with unhappiness, one must accept it. . .those accustomed to endure misery learn to scorn it."[43] Tran Van Tung concludes: "The practice of Buddhism leads to pessimism and fatalism, which have entered into the conscience of the Vietnamese people. Folk songs, epics, and poetry are a moving expression of this scorn for earthly life."[44] A Vietnamese scholar, Nguyen Kyen, wrote:

> Thinking on life is quick to bring disenchantment! To what avail hope. *Why plan when all depends on the heavens? The greater the ambition the harder the deception*, more bitter the regret. We win the battle but lose the cause. *Life*

> *is endless disillusion.* Success, failure, gain, loss, are not worth a jug of wine. Good, evil, praise, blame, all die with a laugh. Leaning on my pillow by the screen I dream of becoming a butterfly. The autumn wind blows cold; the leaves of the jacaranda fall [emphasis mine].[45]

This view of existence has also been noted by Western observers. Hammer writes: Their profound fatalism has been nourished by religion and by personal experience. The average Vietnamese has never been in a position to control his own environment, and he is reluctant to act on his own initiative.'"[46]

It is important to note, however, that the Vietnamese do not entirely accept fate with abject resignation. They attempt to shape their environment to the extent that past experience or history has indicated such efforts may be worthwhile. In this regard Tran Van Tung notes that in the Vietnamese struggle "against misfortune and destiny, courage, patience, and goodwill are indispensible."[47]

The significance of this cultural characteristic for the PLA is apparent. The PLA soldier, finding himself in a main force unit through a process over which he had little control, tended to become willing to endure extreme hardship and danger for the accomplishment of PLA objectives. A prisoner from a main force unit, asked about his chance of survival if he had not been captured, stated: "The fierce fighting would bring death to most people. It would be difficult to survive and come home to the family if the war went on. . . ."[48]

Linked to their profound fatalism is the Vietnamese view of death. Hammer notes:

> Death is an exit which they not infrequently choose for themselves deliberately, sometimes for motives of honor but sometimes also for surprisingly futile reasons. Life is all the less valued because of the Vietnamese belief in the survival of the individual after death; and continued contact with the family.[49]

In war the Vietnamese revere death in defense of the nation as heroic and honored, a tradition which the PLA reinforced.[50]

Death, viewed differently than in the West, appeared to be less of a deterrent to the PLA soldier. One POW stated: "About 30 percent of the men in the unit were afraid of death and the continuation of war. Seventy percent of the men weren't afraid. They kept on fighting to the end without hesitation."[51]

In summary, the group orientation, the fatalism, the view of death, and the great potential for nationalism within the Vietnamese culture provided valuable raw material from which the PLA cadre formed malleable groups and eventually first-rate combat units.

NOTES

1. John and Marvis Biesanz, *Modern Society* (Englewood Cliffs, N.J.: Prentice-Hall, 1968), pp. 91-93.

2. Alexander L. George, "Primary Groups, Organization and Military Performance," in *The Study of Leadership* (West Point, N.Y.: USMA Printing Plant, 1973), vol. 11, pp. 20-22, 20-23.

3. Ibid., p. 8. For similar propositions see Morris Janowitz and R. Little, *Sociology and the Military Establishment* (New York: Russell Sage Foundation, 1965), or E. A. Shils and M. Janowitz, "Cohesion and Disintegration in the Wehrmacht," *Public Opinion Quarterly* 12 (2), 1948.

4. Cited in George, "Primary Groups," p. 9. George points out that this proposition has been supported by other investigators as well. For example, see Alexander George, *The Chinese Communist Army in Action* (New York: Columbia University Press, 1967); Shils and Janowitz, "Cohesion and Disintegration"; Janowitz and Little, *Sociology*.

5. This definition of nationalism and the accompanying criteria for measuring its potential are taken largely from a series of lectures delivered by Professor Richard Cottam at the University of Pittsburgh during the summer of 1969. In arriving at this definition Professor Cottam built on the ideas of Emerson and Kohn. See, for example, Hans Kohn, *Nationalism, Its Meaning and History*, rev. ed. (Princeton, N.J.: Van Nostrand, 1965), and Rupert Emerson, *From Empire to Nation* (Boston: The Beacon Press, 1967). On p. 104, Emerson states that Vietnam fulfills the classic (that is, French) criteria for nationalism.

6. The values assigned to Vietnam may also be arbitrarily compared to similar criteria for such countries as India and the Soviet Union in order to gain an appreciation of the strong potential for nationalism shared by the Vietnamese.

7. Roy Jumper and Marjorie Weiner Normand, "Vietnam: The Historical Background," in Marvin E. Gettleman, *Viet Nam: History, Documents and Opinion* (New York: Fawcett, 1968), pp. 10-28.

8. Vietnamese historical archives are virtually nonexistent. During the thousand-year struggle with the Chinese such records were deliberately destroyed or carried off to China to prevent the Vietnamese from studying their history. More archives were destroyed during the widespread turmoil following World War II. Contemporary records are also lacking. Many valuable papers were destroyed in the burning of the presidential palace during the coup which overthrew Ngo Dinh Diem in 1963. Of historical records that are available many are of suspect validity because they have been falsified or altered for personal and ideological reasons. For more details see Ellen Hammer, *Vietnam, Yesterday and Today* (New York: Holt, Rinehart and Winston, 1966), pp. 54-55.

9. For an example of Vietnamese legends that contribute to the formation of a Vietnamese identity see Do Van Minh, *Viet Nam: Where East and West Meet* (Milano: Casa Editrice, n.d.).

10. Hammer, *Vietnam*, p. 73.

11. Ibid., pp. 220-21.

12. Ibid.

13. During the spring of 1967, the First Air Cavalry Division captured such a PLA traveling troupe complete with scripts and American uniforms to assist in portraying "American Imperialism."

14. For example, see "A Brief Chronology of Momentous Facts and Events in the History of the Democratic Republic of Vietnam," in Document no. 84, *Viet Nam Documents and Research Notes*, American Embassy, Saigon, Vietnam.

15. Douglas Pike, *Viet Cong, The Organization and Techniques* (Cambridge, Mass.: The MIT Press, 1966), p. 374.

16. Ibid., p. 2.

17. Ann Crawford, *Customs and Culture of Vietnam* (Tokyo: Tuttle, 1968), pp. 55-61.

18. The PLA recognized the potential difficulties of communication and group cohesion which might have arisen with these minority "nationalities" and, therefore, organized ethnically pure minority units within the PLA with their own semiautonomous areas for military operations under ethnic leaders. This did not apply to the overseas Chinese who resided primarily in Cholon and from whom large sums of money were extracted.

19. Hammer, *Vietnam*, p. 39.

20. Regional differences such as into-national dialect, pronunciation, preparation of food, and general attitudes have on occasion been bases for conflict between northerners and southerners in the PLA.

21. Pike, *Viet Cong*, pp. 5-6.

22. Based on conversations with Major William Brady, CGSC Fort Leavenworth, Kansas, January through March 1974.

23. Interview K-1, p. 28.

24. Gettleman, *Viet Nam*, pp. 36-40.

25. Hammer, *Vietnam*, pp. 220-21.

26. Ibid.

27. Ibid., p. 1. For an excellent discussion of recent Vietnamese nationalism and its origins see John T. McAlister, Jr., *Viet Nam: The Origins of Revolution* (New York: Knopf, 1969).

28. Paul Mus, "Vietnam: A Nation Off Balance," *Yale Review* 41 (1952), p. 528.

29. Villages were usually populated by extended kinship systems which traditionally have had a close relationship with the notion of a Vietnamese nation. Historically, when new villages were established as the Vietnamese pushed their way south, the inhabitants would apply to the emperor for a village name. When the imperial document giving the village its name was received, the villagers built a *dinh* or house for the document. This identity with their village strengthened Vietnamese awareness of themselves as a people. For further details, see Gerald C. Hickey, *Village in Vietnam* (New Haven, Conn.: Yale University Press, 1966), pp. 5-6.

30. Pike, *Viet Cong*, p. 6.

31. This system was the preferred method of obtaining main force recruits but it was not always possible to follow, especially after the war intensified in 1965. This forced the VC to draft directly to main force units, creating problems which often required increased surveillance and unit isolation. See Michael C. Conley, *The Communist Insurgent Infrastructure in South Vietnam* (Washington, D.C.: Center for Research in Social Systems, American University, 1966), p. 118.

32. Pike, *Viet Cong*, p. 238.

33. The term *subject* is used in the sense of Almond and Verba, that is, a citizen for whom things are done by the government and over which he has no influence. See Gabriel A. Almond and S. Verba, *The Civic Culture* (Boston: Little, Brown, 1965).

34. Tran Van Tung, *Vietnam* (New York: Praeger, 1959), pp. 53-54.

35. The techniques of control and motivation used by the cadre-leader, particularly at the platoon and squad level, as surrogate father will be discussed in Chapter 5.

36. Comments made during informal discussion in a seminar on insurgency at the Command and General Staff College, Fort Leavenworth, Kansas, academic year 1973-1974.

37. John T. McAlister, Jr., and Paul Mus, *The Vietnamese and Their Revolution* (New York: Harper and Row, 1970), p. 96.

38. These techniques will be fully discussed in Chapter 5.

39. Konrad Kellen, *Conversations with NVA and VC Soldiers: A Study of Enemy Motivation and Morale*, RM-6131-1-ISA/ARPA (Santa Monica, Calif.: RAND Corporation, September 1970), p. 31.

40. Ibid., p. 32.

41. Some soldiers, in spite of high merit, found themselves limited in advancement because they were of the wrong class background or because of a relative on the GVN side.

42. Interview K-37, RAND Corporation, Santa Monica, Calif., p. 23.

43. Tran Van Tung, *Vietnam*, p. 30.

44. Ibid., p. 30.

45. Quoted in ibid., pp. 42-43.

46. Hammer, *Vietnam*, pp. 27-28.

47. Tran Van Tung, *Vietnam*, p. 30.

48. Interview K-8, p. 25.

49. Hammer, *Vietnam*, pp. 27-28.

50. Ibid., p. 28.

51. Interview K-9, p. 25.

5

The Cadre-Leader

THE CADRE-LEADER AS AGENT OF THE PARTY

Central to Lenin's thought was the idea that given an organization of professional revolutionaries the world could be transformed. In this sense the cadre are the organizational and ideological core of professional revolutionaries within the Communist party. The creation of party cadre is a basic job of all party organizations. It is accomplished by a politization process which lasts years, from recruitment through development into a highly sophisticated and "deployable" agent of the party.[1] The years invested in developing such men are not regarded as excessive since the cadre form the indispensible base from which the "organizational weapon" is guided. Absolute devotion and loyalty to the party and a desire to maintain the "closest possible contact with the masses" are inculcated by various techniques of "ideological indoctrination," and "political combat."[2] A party organization based on cadre, according to Philip Selznick,

> . . .is a highly manipulatable skeleton organization of trained agents; it is sustained by political combat and is linked to the mass movement as its members become leaders of wide groups. . . .[3]

The term *cadre* comes from the French where it means "frame." In this respect it is often referred to in military terms to mean the battle-hardened and experienced troops around whom a mass force can be developed.[4]

In the Vietcong, these two definitions were combined to form what Douglas Pike calls the secret of Vietcong successes—an organization consisting of cadre superbly suited to their task.[5] Some details of this combination of political and military skills are supplied by Professor J. J. Zasloff. The Vietminh cadre who were veterans of the struggle against the French made unnecessary the long process of recruitment and training needed to produce new cadre.[6] Vietminh "stay behind" cadre constituted a "ready made apparatus" which spread throughout South Vietnam, and was supplemented by some 30,000 "regroupees" returned from North Vietnam. Together they formed the "steel frame" which became known as the National Liberation Front.[7] These Vietminh veterans provided two essential skills necessary to the Vietcong organization. First, most were professional soldiers experienced in the many skills necessary to a military organization. Second, they were dedicated to the party and its ideology, approaching their tasks with a missionary zeal.[8]

The importance of cadre is made clear by the following quotation from the official PLA newspaper, *Quan Doi Nhan Dan.*

> In the revolutionary struggle, we all know that, after determining a correct political line, the decisive condition for success of this line is to organize its implementation in which the key problem is building a contingent of *cadres firm and strong in every respect.* Since cadres are a core force having a decisive nature in organizing and the implementation of every line and policy of the Party. . .our Party and President Ho, in the leadership over the (PLA), have shown their great concern over building a contingent of cadres, considering it as "*precious capital*" *of the Party* and the core of revolutionary forces. President Ho taught: "*The success or failure of everything depends on whether cadres are good or bad*" *and "with good cadres, everything can be done.*"
>
> In our armed forces, a cadre must educate and train fighters in every respect on the basis of the Party's line, policy, guideline, viewpoint, and thoughts while commanding them in combat and in every task.
>
> Although fighters are skillful, the armed forces cannot develop their revolutionary nature and combativity if cadres

> are not good organizers and commanders. . . . As taught by President Ho, "*Cadres are the root of all tasks. Cadre training is a basic task of the Party which must feed and teach cadres, just like a gardener who cultivates precious trees. It is necessary to esteem talented people, cadres, and every person who is useful to our common work. . . .*[9]"

During the 1965-1967 period, cadre of this quality manned the PLA.[10] At the heart of the cadre's role, especially at the squad, platoon, and company level, was the indoctrination and direction of the soldiers in accordance with party goals. To accomplish these tasks he was provided with a variety of approved techniques and mechanisms. Using George's model, these techniques and mechanisms can be categorized under the headings of "comradely leader" and the "morale informants system."[11]

THE CADRE-LEADER AS A COMRADELY LEADER

When building "comradely relations" the cadre-leader functioned in the role of a leader concerned with his men's welfare. He attempted to create a favorable first impression on new soldiers. He made every effort to set the example of the "good communist soldier." He intervened where possible for the men. He attempted to satisfy their material needs, welfare problems, their complaints against the military leaders, and other problems. In combat, according to a PLA document, the cadre provided

> Solace for hardships, encouragement for carrying out the offensive. When there are casualties among his associates, the soldier's resentment should be stimulated. During a fierce battle we will use stimulating slogans. . . . After combat, we will give rewards. . . .[12]

Another PLA document notes:

> The cadres should contribute their constructive opinion, argue logically with the unit members, and solve all difficulties for their subordinates. . . .The cadre should set a good example during training and study sessions. Only

> in this manner can they motivate and urge the sections within their unit to apply themselves.[13]

Fulfilling the role of comradely leader was a difficult task. Douglas Pike points out that the party ultimately held the cadre responsible for all "failure and error."[14] Another document from this PLA battalion, evaluating cadre shortcomings in 1966, noted:

> *Short-comings.* In the case of the cadres, a small number were guilty of "bureaucratism" (*quan lieu*) in their dealings with their subordinates; for example, they were hot-tempered, shouted at their subordinates, and treated their subordinates without affection. They provided shallow leadership. For example, sometimes they did not immediately indoctrinate the men who showed signs of demoralization and discouragement in their daily activities, and this was why in the past a number of comrades have deserted.[15]

THE CADRE-LEADER AS OPERATOR OF THE MORALE INFORMANTS SYSTEM

As operator of the morale informants system the cadre-leader was concerned with resocialization, attitude change, and mission accomplishment. At the highest level, the cadre's goal was to remold the man into a "good communist soldier" with the potential of someday becoming a cadre-leader. At the lowest level his goal was to mobilize the uncommitted, even if only temporarily.

In discussing the Chinese PLA George states that surveillance, self-examination, mutual criticism, and self-criticism are interrelated processes aimed at "thought reform" and control.[16] These processes had two general results: first, they remolded some soldiers' thoughts along the desired ideological lines; second, even when they failed in ideological remolding they usually succeeded in suppressing undesirable activity while mobilizing and directing the PLA soldiers along desired paths. George states that two types of soldiers generally emerged: first, the converted, who were referred to as "beet red" or red throughout; second, those who maintained the facade of conversion, called "radishes" or red on the outside only.[17]

In examining why some soldiers became beets, George, referring to a study conducted by Hiniker, states:

> . . ."internalization" [genuine attitude change] by means of "forced compliance" [the morale informants system] seems to depend heavily upon the prestige of the authority in the subject and may indeed require an intermediary stage of personal identification with authority to become operative. We suggest that an intermediary state of this kind is precisely what the communists attempted to provide in the institution of the company political officer and lower-level politicized cadres. We noted (earlier) the ways in which company political officers, often successfully, attempted to develop the friendship and respect of soldiers under them in order to facilitate their indoctrination and assimilation into the . . . system.[18]

As the quotation suggests, the cadre linked the authority structure to the soldier. Further, George shows how the Chinese PLA cadre served both as comradely leader and implementer of the morale informants system.

> The company political officer spent considerable time observing the men working, marching, eating, and resting. He often dropped in on squad and platoon meetings; and, insofar as time permitted, he circulated throughout the company talking to soldiers individually or in groups in order to find out how they were getting along. The political officer's opportunities for direct contact and surveillance were limited, however, especially during combat. Moreover, he knew that the men were likely to control expression of disapproval attitudes when he was around to a considerable extent; therefore, he relied upon platoon, squad and group leaders for moral surveillance.[19]

Within the Vietnamese PLA, cadre use of the morale informants system was readily apparent. The cadre administered surveillance, criticism sessions, and other similar control techniques. For example, a squad cadre, according to one interviewee, must keep "watch over the three man cells and to report once every three days to the Party leaders about the situation and activities in the cell." A cadre-leader who was a POW stated:

> . . .the fighters were watched very closely from the lowest squad level up. Therefore, we had all the information about their behavior and performance from one day to another. By watching them directly and indirectly, that is to say, by following up the reports made on them, we cadres knew whether a man had really good morale or not. Even if a man pretended or hid something, he would eventually be found out, no matter how well he pretended or how carefully he hid. Take, for example, the case of a fighter having an affair with a female: in such case, he'd of course try to conceal it from other people as much as possible, but finally he'd be found out. The same thing can be said of any weakness—how can it possibly be hidden all the time.[20]

An important element of the morale informants system was the PLA indoctrination effort. The PLA themes of indoctrination, as they related to the 1965-1967 period in South Vietnam, can be divided into two categories. One, summarizing what the PLA was fighting against, included "American imperialism," the feudalism of the successors to the Diem regime, and the "Vietnamese upper class." A second category involved what they were fighting for: "democracy; reunification; and an end to poverty [and] injustice, and [achievement of] peace."[21] These themes provided the basis against which activity within PLA squads, platoons, and companies was measured. For example, exhortations to "labor for the revolution" and to "conserve" were used as the basis for individual and unit production and ration quotas. An individual's action could thus be evaluated within the context of the morale informants system as to whether it promoted or hindered progress toward the accomplishment of PLA goals.

THE LINKAGE PROCESS

How well the cadre-leader performed the roles of operator of the morale informants system, comradely leader, and agent of the party determined his success in accomplishing four functions necessary for the unit to fight and endure for an extended period of time.[22]

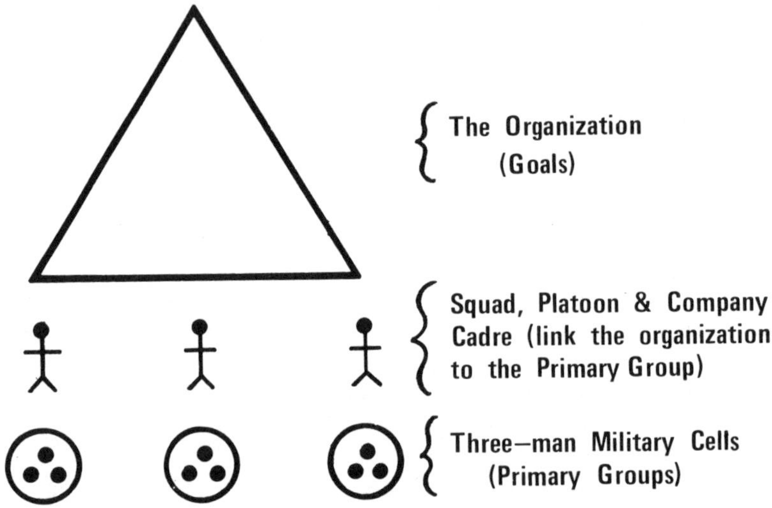

Figure 4. The Linkage Function of the Cadre.

These four functions were: (1) the transmission of organization (party) goals to the military unit; (2) leading the unit in accomplishing party goals; (3) the surveillance of individual and group performance and attitudes to detect deviance and ensure compliance with party goals; and (4) creating "good communist soldiers" through indoctrination.

THE TRANSMISSION OF PARTY GOALS

Cadre were reminded repeatedly of the importance of maintaining the party's dominance in the military organization:

> Party leadership is the cause of, and the decisive factor in, the birth, growth, and victories of our Army: leading the

> Army is our Party's historic task; and constantly and firmly maintaining and thoroughly implementing Party leadership is the highest principle of directing all activities of our Army. . . .[23]

Party documents recognize "military science" as having its own principles and techniques. However, these were not accorded autonomous status in determining battlefield objectives.

> Military science and techniques are a branch of natural sciences. But their building and application must be closely associated with the Army's political objectives, building and combat line, and class nature. Thus, in building and applying our military science and techniques, we must be guided by our Party's political line and goals. . . .[24]

Specific guidance on how to achieve party guidance over the military was given to the cadre:

> In our armed forces, a cadre must educate and train fighters in every respect on the basis of the Party's line, policy, guideline, viewpoint, and thoughts while commanding them in combat and in every task.[25]

An illustration of the cadre's role in transmitting party goals to the "fighter" is provided in the following excerpt from an interview with a rallier:

Question: What were the responsibilities of your unit's political officer?

Answer: The political officer was responsible for the Party activities, security, transportation, and so on.

Question: Then who was responsible for the political education and morale problems in your unit?

Answer: The political officer was also responsible for that.

Question: Did the political officer in your unit do his work well?

Answer: He organized his work very well. He made careful plans for our work.

Question: Did you think the work of the political officer was important?

Answer: Yes, I did. I can say that the political officer was the mother of the unit. He took care of everything for us.

Question: What made you think that his job was important?

Answer: *He represented the Party in the company and saw to it that the Party policies were carried out properly. And he was always with the unit to make changes in the policies to fit them into the unit. He was the Party authority in the unit.* It was his job to make the unit strong. And he was to blame if the unit was weak. And I am sure that the higher cadres would blame him for my rally [emphasis mine].[26]

THE LEADERSHIP FUNCTION

Through his role as comradely leader the PLA cadre-leader influenced the individuals within the primary group toward the accomplishment of organizational goals, an exceedingly complex task, especially in combat. Amitai Etzioni points out that organizational goals are often in direct conflict with the individual's best interest in a combat situation.[27] For example, the individual's need for safety from enemy fire is in direct conflict with the organizational requirement to advance toward an enemy position and defeat it. The function of the cadre-leader was to bring about congruence between the requirements of the organization and the needs of the individual. The cadre-leader had to create an internalized discipline within the soldier to help him overcome his fear and expose himself to enemy fire. To accomplish this, the PLA cadre-leader manipulated and accommodated individual needs by developing a group whose norms were strongly congruent with organizational objectives. In other words, the PLA required the soldier to pursue PLA objectives in satisfying his individual needs. Shils and Janowitz describe a similar situation in which Wehrmacht units fought to the end so long as the soldier's major needs were met by the group, resulting in control of his behavior by group expectations.[28]

The influence of such group norms can be exceedingly strong, especially in combat situations. As previously noted, one PLA cadre-leader, a prisoner, claimed that "70 percent of the men in his unit weren't afraid of death but fought to the end without hesitation."[29] In threatening combat situations characterized by uncertainty and ambiguity, the soldier was inclined to let group norms dictate his behavior in return for group satisfaction of security, affection, and other needs. The task of the cadre-leader was then to ensure that those norms which emerged in the group were congruent with organizational goals. Numerous leadership studies have shown that leaders not designated by the organization often emerge, thereby reducing or nullifying the influence of the "leader" appointed by the organization.[30] To preclude this, a variety of leadership techniques were employed by the PLA to ensure that the appointed cadre-leader was the actual or emergent leader as well, thus reducing the possibility that autonomous subgroups would evolve.

The PLA psychological approaches, along with those of the Chinese PLA, may be the most sophisticated leadership techniques in use by any army in the world today. The following discussion describes the approach used by PLA cadre-leaders to manipulate individual and small group behavior.

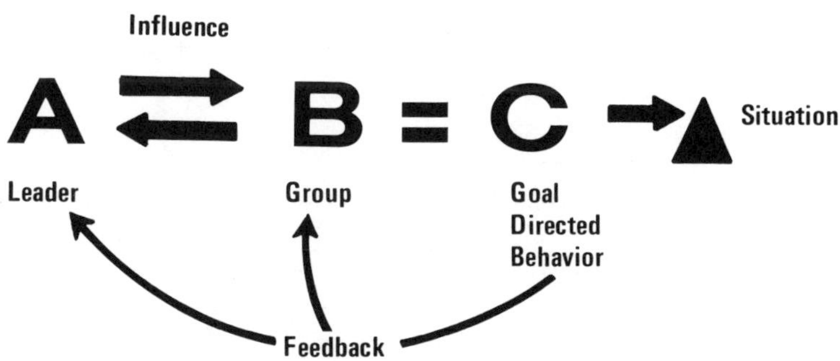

Figure 5. The Leadership Model.

Leadership (see Figure 5) may be defined as the phenomenon that occurs when the influence of A (the leader) causes

B (the group) to perform C (goal-directed behavior) when B would not have performed C had it not been for the influence of A.[31]

Interaction between the leader (A) and the group (B) is signified by the two arrows and indicates the exercise of influence upon which the leader relies to influence the group. The behavior depicted by C is mission-oriented activity desired by the leader, as the agent of the organization, and performed by the group. Feedback enables the leader and the group to adjust their activities over time as the situation changes.

There are several bases of cadre-leader power which permit him to influence small PLA units in accordance with party objectives. These may be placed into five categories evident at the squad, platoon, and company level: (a) reward power; (b) coercive power; (c) legitimate power; (d) referent power; and (e) expert power.[32]

REWARD POWER AND COERCIVE POWER

Reward power may be defined as the ability to exert influence in interpersonal relationships based upon the ability to reward. Likewise, coercive power is defined as influence whose basis is the ability to punish. Within the PLA, the cadre-leader did not have wide latitude to reward materially or to coerce individual soldiers. Resources for rewards such as money or luxurious rest-and-recuperation vacations were nonexistent for the PLA. Similarly, physical punishment and incarceration were seldom relied upon. In extraordinary circumstances a deviant soldier might be assigned to a "reeducation" camp for two to six months. However, this was not perceived as imprisonment. Instead, as the name implies, the soldier was reeducated along the lines of "correct thought." Upon the successful completion of his reeducation, he could, theoretically, rejoin a unit without bias.

As practiced within the PLA, reward and punishment were almost always related to the soldier's relationship with the primary group. The cadre's ability to regulate the group's acceptance or rejection of the individual was a source of tremendous power in view of Vietnamese dependence on group relationships.

Among the most successful reward techniques were the individual and unit awards and commendations. Through these awards the individual was granted group recognition and esteem. Punishment was generally a group sanction entailing loss of face through a criticism session in which the violation of group or organizational discipline was publicly discussed, causing shame for the individual.

Rewards and punishment were always related to group (PLA) norms. For example, a soldier criticized for being lazy might be cited as lacking in "commitment," or as guilty of "rightist thoughts" or "freedomism" (acting for one's own convenience), whereas an exemplary achievement might be cited as a commendable example of "correct thought," "virtue," and "revolutionary spirit."

The following excerpt from a platoon party chapter meeting illustrates how "Comrade Phuoc" was rewarded and how "Comrade Minh" was punished:

> COMMENDATIONS AND DISCIPLINARY MEASURES
>
> a. The. . .Group Chapter decided to have the shining example of Comrade Phuoc studied by the entire PRP Youth Group.
> b. Comrade Minh was purged from the Group and put on probation for six months—if he made progress during this period he would be readmitted in the Group. He was disciplined for stealing two cans of rice from the unit, and for not observing the discipline of the unit. For example, he left the unit without permission, and when he was criticized by the collectivity he refused to admit his error.[33]

The PLA constantly shifted its organizational goals to meet the changing tactical situation. "Emulation" campaigns within PLA ranks were designed to keep the soldier informed of the group's specific goals, providing a basis for measuring his contribution. For example, from September 1966 through the autumn of 1967, the PLA mounted a program entitled "Troop Training and Combat Competition—An Emulation Plan," designed to intensify its combat efforts against U.S.

forces.[34] The program spelled out how PLA units could earn "good" or "fair" ratings. A PLA company was required to destroy one U.S. platoon or two "puppet" (ARVN) platoons for a "good" rating or "annihilate two U.S. squads or one puppet platoon" for a "fair" rating. In addition, an emulation "Heroic American Killing Company" flag was to be awarded every six months to the best company. The same program also prescribed that titles such as "Heroic American Killer," "Heroic Aircraft Annihilator," "Heroic Armored Vehicle Annihilator," and "Assault Hero" were to be awarded to appropriate individuals. A private discussing this program stated that his squad leader:

> . . .killed 22 American soldiers in that battle and was elected "Valiant American Killer." He was also sent to attend the "Valiant American Killers Congress" in March 1966.[35]

Each soldier's performance in achieving group goals was routinely scrutinized by the cadre-leader. The following excerpt from an interview of a main force corporal describes how rewards and punishments were mediated through the group.

Question: Do you think that criticism and self-criticism is good or bad?

Answer: It is good. If there wasn't that system, the fighters' rank would immediately disintegrate. The criticism and self-criticism is a formidable weapon. I can tell you if there wasn't the criticism and self-criticism the fighters would all desert in one day. There was criticism and self-criticism every day. After a working day, a self-criticism session was immediately held at night. There were thirty nightly sessions a month. The criticism and self-criticism was part of the rules and regulations. It couldn't be missed. . . .

Question: What did the other men say? (about whether it was fair).

Answer: They also thought that the system was correct because *it helped the cadres to make a difference be-*

tween good and bad men. Those who had combat achievements or who did good work were commended, and those who had shortcomings were criticized. This was the guideline of the leadership. [emphasis mine].

Question: Do you think the criticism helped to make your unit a good one?

Answer: Oh yes! The courage and the exploits of the men all originated with the criticism and self-criticism. Everyone made it a point of honor not to be criticized.[36]

While a system of reward and punishment is in the toolchest of almost every army, the PLA has refined it to the point where it has become a major motivational and control technique for reinforcing desired behavior and discouraging deviant behavior.

LEGITIMATE POWER

Legitimate power may be defined as compliance with orders because of attitudes or beliefs which have their basis in a feeling of internalized "oughtness" that in turn is based on learned cultural values. In this respect the legitimate power of the PLA cadre appears to rest on Vietnamese cultural values which include fatalism, respect for age, and nationalism. In addition, the fact that PLA cadre were appointed by an organization which had achieved some degree of legitimacy through its record of effective government in North Vietnam and in certain areas of South Vietnam added to their power.

In 1965, almost all men between the ages of eighteen and thirty-five living in areas controlled by the VC were available for drafting into the PLA.[37] The fatalistic draftee, as noted earlier, was likely to believe that he was in the PLA through a process over which he had little control. He tended to respond to the situation with courage, patience, and goodwill.

Cadre reliance upon legitimate power, especially for controlling draftees, tended to be greater during the earlier period of a soldier's service. Later, the bases of cadre power shifted away from legitimate power. Interviews of draftees and volunteers with long periods of service suggested that both had similar motivation and no distinction can be made in their motivation whether they were drafted or had volunteered.

Other Vietnamese cultural values which contributed greatly toward cadre legitimate power were respect and deference for "elders." The transference of this respect to the PLA cadre was a common occurrence among PLA soldiers in their teens and twenties. In a very real sense the cadre became a "respected elder" or surrogate father, fulfilling an important authority role. Numerous interviewees referred to the unit political officer in particular as the "mother" or parent to the group, guiding and caring for it in every way, as the following excerpts illustrate:

> *The troops in a unit considered the political officer as their mother.* This cadre always tried to get the opinion of the troops in order to know and settle the disputes among the soldiers. . . and saw to it the unit was unified. Besides the ideological training, the political cadres also promoted the fighting spirit of the soldiers and took charge of their subsistence, i.e. food and drink, etc. *Because of this devotion, by the political officer, the troops in a unit liked and respected the political officer very much. Due to such respect and confidence, the troops could always overcome the difficulties in the fighting, as well as in the daily work, carry out thoroughly the orders of the cadres, and achieve good results for the unit. . .*[emphasis mine].[38]

The following excerpt indicates how cadre authority was often interpreted in a parent-child context by the "fighters":

> . . .If the fighters said that the squad leader was harsh to them, he [the political officer] would explain to them that the squad leader had to be harsh because *he wanted them to be good and respectable. He also reminded them that people often said, "When one loved his children, he gave them spankings, and when one hated his children, he let them play (a Vietnamese proverb)."* To me, his explanation about the squad leader's behavior was very correct and very pleasing [emphasis mine].[39]

The fact that the Vietcong were identified as fighting for Vietnamese independence also contributed to cadre legitimacy.

As the successors to the Vietminh, who struggled against the French, many Vietnamese saw the PLA as merely the latest organization to take up the traditional banner of Vietnamese nationalism. One PLA prisoner stated:

> I think that the people here support the Liberation Front because, in previous times, they had fought during the First Resistance against the French. They want Vietnam to be independent and reunified. The continue to support the Front because Vietnam is not yet independent and reunified.[40]

An unusually articulate PLA private was more specific in blaming the war on the Americans and portraying the PLA cause as one of Vietnamese independence. He stated:

> . . .we were told to fight the American aggressors who had sabotaged the Geneva Agreement. If the Americans didn't enter Vietnam after this agreement, Vietnam would have been reunified by general elections and would have enjoyed peace a long time ago. . . .According to the VC, general elections and peace would follow the battle of Dien Bien Phu and the Geneva Agreement within two years. But, after this fixed period, the Americans brought back Ngo Dinh Diem from the U.S. and used the former soldiers of the French to sabotage the Geneva Agreement. Ngo Dinh Diem has drafted many young men into his army. In my opinion, Vietnam is only one country and cannot be divided into two parts. The Vietnamese people should not fight against each other. If the Americans didn't pour their troops into Vietnam, perhaps this war wouldn't have happened and the people might have elected Mr. Ho Chi Minh or somebody else as their leader or their president. I repeat that this war has been started by the Americans alone, but I don't fully approve of the war policies of the VC, and of the Communist Chinese.[41]

Finally, the mere fact that a cadre-leader holds a decision-making position in a legitimate organization such as the PLA means individuals will respond to his direction in much the same manner that workmen respond to a foreman, or a parish responds to a priest. Cadre-leaders appointed by a legitimate organization tend automatically to benefit from a "halo" effect. A fighter stated that one reason why PLA soldiers fought for the PLA was:

> . . .because they wanted their families to be honored by the Front and by the other villagers. For example, if there was a Front meeting in the village, those who had children in the Front troops would be invited to sit in the front seats. The villagers liked to be honored that way.[42]

A more common source of legitimacy for the PLA and the cadre-leader is evident in the following excerpt from a regroupee prisoner:

> . . .A unit is strong when its men recognize that *their duty is to defend their own home area.* When they see that houses have been set afire, cattle and other properties being destroyed and damaged, and when they are continuously stimulated and lectured on their duty, they can't help recognize their obligation to fight with fervor and courage. . . [emphasis mine].[43]

The PLA, like the Chinese PLA before it, made great efforts to engender support from the people.

> . . .the men in the unit would always help the villagers at all types of work: for example, if the villagers were working on their land, or fishing, or digging ditches, building dams, etc., we would come and give them a hand. This is why there was never any conflict between the people and us. If we had caused them trouble, either damaging their land or taking everything from them, then they would have reason to complain about us, but that wasn't the case. On the contrary, *wherever and whenever we arrived we gave help to the people; they were thus very appreciative of our sincerity and loved us very much.* . .[emphasis mine].[44] Our motto is: The Army is welcomed wherever it goes, and loved wherever it stays. We will never touch even a piece of thread that belongs to the people. What belongs to the people remains theirs, and if by mistake it's damaged by us, it will be compensated.

> What has been taken away from them by mistake will also be returned or paid for. We stand for helping the people and loving them. However we can be useful, we will do it for them. We are to promote good relationships between the people and the Army.[45]

EXPERT POWER

A person's compliance with a cadre's orders because the cadre is perceived as having superior knowledge or ability may be defined as expert power. In combat especially, perceived expertise in coping with the situation is a source of tremendous power for the successful leader. The demonstrated ability to provide a successful tactical plan, to arrange for an adjust supporting artillery fires, to demonstrate proficiency with weapons and map reading, and to provide for medical evacuation, medical care, and adequate supplies is a proven source of power.

The PLA recognized the importance of the cadre-leader being an expertly qualified person. As an agent of the party, the cadre-leader was required to be of high "ethical virtue." A second requisite was "talent" in military affairs, as the following quotation indicates:

> To have an excellent cadre contingent serving as nucleus of the armed forces, the Party set forth *an ethical and talent criteria. . . .* Morality and ability are two fundamental qualities that all our officers must have. He who lacks one of these qualities cannot become an officer. President Ho taught us: "*He who has talents must have virtues. He who has talents, but has no virtues. . .does harm to the country. He who has virtues, but no talents resembles a Buddha who, staying in the pagoda, does nobody any good.*"[46]

Evidence is plentiful that PLA soldiers saw their cadre-leaders as "men of steel," capable of expertly performing, for the benefit of the group, all tasks demanded by the situation. The following excerpts show the soldiers' awe and respect for the cadre. One soldier, a cadre-leader himself, stated:

> . . .all of them had been trained in specialized schools, and matured in actual combat. Therefore all of them were worthy of confidence and admiration. Of course I had confidence in them. I could see they were men of good experience; I felt proud of them. . . .

In large part he saw his influence with the fighters dependent upon his performance in combat.

> . . .I also participated in direct combat side-by-side with the fighters. You see, the thing was, *if you were a good fighter, then you would be listened to by others; moreover, if you treated others well, and if you were a good fighter also, then your work as a political cadre was already half over.* A political cadre should be the one who acts first and then speaks [emphasis mine].[47]

Successful decisions reflecting expertise also contributed to cadre influence.

> I as well as other men had much confidence in the commanding cadres as before because they were very clear-sighted. *For example, the commanding cadres ordered the unit to move to another place just a day or half-day before the old campsite was bombarded by aircraft and artillery. This made the men have confidence in the cadre* [emphasis mine].[48]

The placement of cadre-leaders in the fore of all group activities, including combat, was costly for the PLA, as the following excerpt indicates:

> We had some regroupee cadres in 1965, but by the end of the year, none was left. They were either wounded or killed in action. *In combat, especially when a fierce battle took place, more cadres had to make an example for the men. They moved from one spot to another to encourage and lead the men in the fighting. As a result, they were often hit* [emphasis mine].[49]

When heavy cadre casualties required the rapid creation of new cadre, lack of expertise reduced cadre influence among the fighters.[50] The following sentiment from a private whose unit suffered heavy cadre losses was common from fighters who saw their cadre as basically unable to lead the group effectively.

> Because of heavy losses, most of the company, platoon, and squad leaders were new replacements. . . .The new cadres weren't as good as the earlier ones. They didn't know much about strategy and had no experience in tactics. In the last months, many fighters showed no respect and less obedience to platoon and squad leaders because most of those people didn't deserve the appointment and were not up to their tasks.[51]

REFERENT POWER

In many cases the most potent source of cadre power was the intense identification of PLA soldiers with their cadre-leaders. Sometimes the cadre-leader approached the status of the charismatic leader who had demonstrated the Weberian quality of "grace" in difficult situations and was expected to do so in the future.

Referent power is derived from the satisfaction gained by the soldier through identification with the cadre. Referent power overlaps other sources of power. For example, reward and expert power contribute to the desire to identify with the cadre-leader. The formation of close identities between the men and the cadre was not left to chance within the PLA. It was official policy that such relationships be developed to their fullest. Cadre were repeatedly told that:

> . . .We must train ourselves into "*simple, modest, diligent, thrifty, honest, selfless, upright, and impartial*" cadres; resolutely eradicate individualism; and insure that in construction and combat and under favorable or difficult circumstances, a correct attitude is constantly maintained toward the enemy, the Party, the people, and our comrades and comrades-in-arms in combat, in work performance, and in other activities.[52]

The following statement by a regroupee prisoner is representative of how the cadre went about establishing close ties with their men:

> As far as relations between the cadres and the fighters were concerned, I can also say that close ties existed between them. Take, for example, the case of some of the fighters becoming ill: often a cadre would take care of the sick fighters. There were also cases of cadres sharing their food and clothing rations with the fighters. *I can tell you that we cadres shared everything with our fighters,* be it a small item, such as a bite of food or a bigger thing, such as money. Very often, with our own money, we went to buy odd things that we shared with the fighters, and vice versa. *If the fighters had the money, it was their turn to spend and share it with us.* There was no case of each one keeping his own possession to himself alone, or hiding it away from others. *The friendship and unity that existed among the cadres and fighters were as close as among the cadres themselves. . . .* [emphasis mine].[53]

The results of such cadre efforts usually produced attitudes similar to those expressed below by a private, first class, discussing one of his cadre-leaders.

> He was the best educated man in the unit. In any circumstance he always succeeded in producing reasonable arguments. *So he won the mind of any person he had to deal with. He was honest and impartial* [emphasis mine].[54]

A portion from another POW interview gives a more comprehensive picture of fighter identity with their cadre:

Question: Describe the cadres in your unit. What kind of persons were they?

Answer: All of the cadres, platoon cadres and company cadres were very nice people. They were well-trained and well-educated. I heard that they were people who formerly had fought against the French. They were well mannered and had adequate fighting experience.

Question: What did you think of them?

Answer: I think they deserved our respect and their positions.

Question: What did the other men say?

Answer: We all respected and obeyed our leaders because, as I told you, they were nice people. They never did anything to hurt the feelings of the men in the unit. *They always lived with us, ate with us, and they understood us very well. We strictly obeyed any order received from them. . . .*

Question: What did you think of the political cadre? How did the other men feel toward them?

Answer: All of the men in our company liked each other. We liked the political cadre very much. *He didn't make us feel that our position was far from his. He talked and joked with us very often; and this made us feel that nothing separated us from the leaders.*

Question: Do you think the political cadre knew everything that was going on in the unit?

Answer: He lived with, and ate with us. Sometimes, when we talked to each other, he came and talked to us too. *I think he knew everything* [emphasis mine].[55]

The role of the unit political officer in particular was important in the maintenance of close cadre-fighter identities, especially during the hardships and dangers of sustained combat. The following excerpt from a PLA rallier interview illustrates the importance of the cadre-leader's role, the identity need he satisfied for the fighters, and the disintegrative effect when this need was not met:

Question: Who was in charge of. . .morale problems in your unit?

Answer: My unit had an excellent political cadre. He was very skillful in convincing people, particularly those who worried about their families. *He used to get in contact with the soldiers in private to advise and comfort them. Everyone liked him and followed his advice. Unfortunately he was transferred to another place, leaving the post vacant for two months. If he had remained longer in my unit, I would have been unable to leave. . . .*

Question: Did the political cadre in your unit do his work well?

Answer: The political cadre in my unit performed his duties perfectly well. Every soldier appreciated and obeyed him. He was very useful to the unit in settling conflicts and in raising the fighters' morale. They strengthened discipline and prevented desertion. *Only during his absence from the unit did many desertions occur, mine included. I think if he left the unit for five or six months without a good replacement, most of the people would desert.* . .his behavior, attitudes and performance always remained the same. *He never lost the heart of any soldier* [emphasis mine].[56]

THE SURVEILLANCE AND CONTROL FUNCTION

A general decrease in the efficiency of motivating fighters accompanied the increase in sustained combat during the latter part of 1966 and 1967. The hardships and dangers of this period diminished the PLA's capability to provide for the soldier's basic needs. In order to maintain control, the cadre-leaders were forced to rely to a much greater degree upon the morale informants system with its pervasive surveillance and control mechanisms. This system appeared to become a more significant factor of cohesion in those units debilitated by combat, hardship, and lack of qualified cadre.[57]

Even as comradely leaders the cadre-leaders relied somewhat upon surveillance and control. Surveillance usually included direct supervision of the fighters at the company level and below. It was built around the three-man military cell (*to tam tam*) and an elaborate reporting system which gave cadre at all levels the capability of monitoring individual and group behavior, attitudes, and morale.[58]

The control of individuals "uncovered" through surveillance was accomplished primarily through two mechanisms: isolation and criticism (self-criticism, *phe binh*, and individual or group criticism, *kiem thao*).

The following excerpts reveal how normal surveillance was intensified when disintegration became a threat to PLA cohesion. Speaking about surveillance of the three-man cell one cadre-leader stated:

> Thanks to the three-man-cells, we were able to keep track of the men's morale. . . .It was hard to tell whose morale was the highest when we were at camp, but during combat, whoever was fearless was the one who had good morale. *As we fought by the three-man-cells, it was easy to check the men's morale. After each battle, the cell leader reported the morale of the men in the cell to the squad leaders who reported to the platoon leaders and so on. As company leader, I got a pretty good picture of the men's fighting spirit.* I never based my judgement on the men's talking manner. Some could be very brave with their tongues, but proved to be cowards during combat [emphasis mine].[59]

A private's perception of the usefulness of the three-man cell was somewhat different but it also recognized the surveillance performed for the organization.

> In my opinion, that three-man-cell system is somewhat useful. When I was lonely in the forest, I could confide to my cell members about my native place, my village. In assignments, men in the same cell could work together, help and follow up one another. In fighting, they fought side by side and covered one another. *That system tied three men together, so as to limit free movement and made desertion difficult. The interdependence in a cell was also good for the command of the unit. Since the men in a cell moved together in combat, they might report the efficiencies and deficiencies of their comrades to the squad leader* [emphasis mine].[60]

Although the fighters generally realized they were continuously watched by their cellmates and cadre, it was normally an accepted and relaxed process, as indicated by the following excerpt.

Question: What did you think of the three-man-cell system?
Answer: The system was very helpful to us cadres. The men could take care of each other by working together. At the end of the day, they could lounge in their hammocks and informally go over their activities during the day and try to check each other's shortcomings. . . .[61]

During the intense combat of the 1966-1967 period, a "security agent" might even be assigned to a unit to observe and report, but this appeared to be the exception.[62] Instead, the PLA often intensified surveillance by party members, as illustrated by the following excerpt:

> It was decided that every three-man-cell should have at least one Party member or Labor Youth because the Party didn't have confidence in the men. It never let three non-Party members be in the same cell. The same principle was applied to every squad.[63]

Direct supervision by the cadre was another means used to intensify surveillance. The following is from an interview with a rallier.

Question: What do you think of the three-man-cell system?
Answer: I think the three-man-cell system helped strengthen the morale of the members because they were able to control one another every hour and even every minute. Thanks to that system the cadres found no difficulty strictly controlling the feelings, thoughts and activities of the fighters. . . .
Question: How did the cadres find out about the state of the morale, fighting spirit and attitudes of the fighters?

Answer: The squad leaders closely watched over the men under their command and reported anything they considered as not normal to their superiors. They were helped in this task by the three-man-cell organization. Sometimes the three-man-cell didn't work. This frequently happened in the last months before I left the unit. All the people were disheartened by events, and no one took care of informing the squad leader of what was happening. My cell leader, quite to the contrary, was too dedicated to his job. He always kept an eye on me and was aware of my thoughts. That's why when I volunteered to go and carry fish sauce for the unit, he, too, requested to participate in the assignment to prevent me from deserting. . . .

Question: Did the fighters in the unit always know who watched them?

Answer: They did. Take my case. I knew quite well that my cell leader watched over me at the request of my squad leader. In order to dissipate their suspicion, for a long while I had been trying to show myself as quite indignant about desertion and always made insulting remarks or criticisms about deserters. Only when they became less suspicious of me could I be given the chance of going on assignment. And that was the only opportunity for me to run away.[64]

Those individuals identified as "lazy," "lacking in commitment," or guilty of "rightist thoughts" or "a lack of an objective point of view," and so on, were subjected to criticism sessions of varying severity, depending upon the gravity of their deviation. Criticism was a psychological technique designed to bring anxiety to the PLA soldier, who was culturally dependent for security upon his relationship with a group. The relief from anxiety produced by the individual's reaffirmation of his intent to comply with group expectations was a strong force for cohesion. As Kellen notes, criticism was "well suited to nip in the bud all possible mental and, subsequently, physical resistance by bringing out not only what some men may worry, or feel antagonistic about, but also by allowing the leadership to apply remedial measures from the beginning."[65]

The desired result of criticism was to guide behavior between fighter and fighter, fighter and cadre, and cadre and cadre in accordance with PLA values and goals. The following excerpt from an interview with a political officer illustrates the usefulness of criticism:

Question: Do you think that the criticism and self-criticism is good or bad? Do you think it is fair?

Answer: I believe that criticism and self-criticism is very good and fair. It is also very necessary for the strength of the unit; it helps to bring about progress within the unit. *Criticism is made in order to improve a person, in order to permit the unit to constantly build itself up in a constructive way; criticism or self-criticism helps a man to get rid of his weaknesses and to be aware of his faults and errors so that he has the chance of improving himself. Self-criticism is a method by which you look into yourself and find your own weaknesses and find ways and means to overcome them.* Criticism and self-criticism are complementary: I make a self-criticism, but there are still one or two of my weak points that I refuse to bring out for review, perhaps out of pride. But my comrades will extract them from me, because I can never conceal them completely, however I try. I make my own self-criticism first, because, as a human being, I have my dignity and honor, and thus I should have a chance to see my own faults and errors before the others. *I believe that criticism and self-criticism help men to mature and to make their actions and thoughts more effective. Criticism and self-criticism also help the men in the unit to achieve the spirit of unity and cooperation.*

Question: What did the other men say about the method of criticism and self-criticism? Were there any men in the unit who didn't approve of it or were against it?

Answer: Generally speaking, everyone approved of criticism and self-criticism. This was because we all recognized that if there was no criticism, the problem—whatever it might be—would be left unsolved among the disputants, and as such, it could only cause annoyance and trouble to the unit as a whole. *Like a small spot of staining oil, you must wipe and clean it out as soon as you spot it, for if you allow it to spread out, then very soon it will cause great damage.* If a small problem isn't solved in its initial stage, it soon will turn into a big issue that can cause great obstruction in the life of the unit. Thus, in many cases, the men in the unit suggested that a criticism session be held if such a session had not yet been called. Apart from the unscheduled criticism sessions, there were regular ones held every month and week: the purpose of those sessions was to permit the men to express their viewpoint, because it was believed that everyone's opinion could be useful and beneficial to the unit as a whole. It was expected that each man would stand up and make his own self-criticism. I believe that everyone in the unit was convinced of the good of those sessions. No one ever expressed disapproval on this point. Of course those who had lots of vices and had committed many faults would naturally feel reluctant to stand up and make a self-criticism, but no one ever denied the usefulness of such a method as a means of self-improvement. . . .

Question: Do you think the criticism helped to improve your unit?

Answer: It was a permanent feature of the activities that contributed to the unit's morale and spirit. . . .

Every unit must hold regular criticism sessions to maintain good standards. The morale of most of the men rose very high after one session of criticism, and the same can be said of the re-orientation classes that were organized every year for the cadres. The men's thoughts and ideology were mobilized, and what had been their weaknesses was given priority, so as to turn them into their strong points instead. . . .

Question: How did you personally feel about the criticism sessions?

Answer: *Holding regular criticism sessions greatly simplified my own work. If there hadn't been any criticism sessions, I would have met with so many difficulties that I would never have been able to manage it. . .* [emphasis mine].[66]

Asked to evaluate the effect of criticism on his unit, a PLA private had this to say: "Yes, it has a good effect. I would compare criticism sessions to a mirror with which I could look at my face. If my face had a stain, I could see it through the mirror in order to clean it up."[67]

Another PLA private was asked if he could explain how criticism sessions "improved bad morale" and corrected "incorrect thoughts." He answered:

> Let me give you an example. For some reason, I do not like you and think you are bad. But I keep this to myself and think that I am right. Now, in a meeting, I criticize you and you defend yourself. Other people will listen to us and join in the discussion. This way will help us know each other better and thus our friendship will be consolidated.[68]

This relief of hostility before it can corrode individual and group behavior is functional for the PLA. Furthermore, the private's comments reveal that he believed high morale and the absence of "incorrect thoughts" come from the sense of security he received from the close relationships within his primary group.[69]

Criticism sessions varied in type and severity. For those generally loyal to group norms, but who had committed minor infractions of group discipline, a mild self-criticism (*phe binh*) session at the three-man cell or squad level was usually sufficient to correct their behavior. However, those guilty of more serious infractions, such as attempted desertion, were subjected to harsher forms of individual and group criticism (*kiem thao*), as one sergeant reported.

> . . .criticism depended on the seriousness of the mistakes. The mistakes could be criticized and decided in the three-man-cell, in the squad, in the platoon but rarely in the company. From the company level up, only those who committed a serious mistake or those who didn't want to recognize their mistakes down in the lower level would be criticized.[70]

While the fighters and cadre-leaders did not always submit to criticism without initial resistance or anger, they generally took comfort in their solidarity with the group upon completion of the experience and accepted criticism as a necessary and beneficial aspect of army life. Even those few who thoroughly disliked the system of criticism because of personal humiliation and shame readily admitted its effectiveness in controlling their behavior.

A final mechanism of control used by the PLA was isolation of units. Isolation was not a normal practice, although it became more common in 1966 and 1967 when frequent moves, insufficient food, inadequate medical care, fatigue, and hazards of combat brought tremendous stress to the PLA organization.[71] The criteria for this action were fairly explicit according to a deputy company commander:

> I knew the personal history of every man in my unit. When a man was sent to my company from a lower unit, his file was sent along. One of our most important jobs was not to let anyone fool us. We had to keep up the men's thoughts all the time. *A normal unit, a company of 120 men, for instance, must have at least 50 percent of the men enthusiastic, 30 percent middle-of-the-road, and 20 percent*

> *unenthusiastic ones. If the percentage of the last category got higher, the unit had to be taken to an isolated place for training. No one would risk taking that unit into combat* [emphasis mine].[72]

The purpose of isolation was to separate the unit from all outside contacts and turn the primary group inward on itself for rebuilding under the strict guidance of the cadre. The unit was situated in a remote area deep in VC territory, if possible. Fighters from the same village and friends made prior to joining the PLA were kept apart to prevent desertion together or the emergence of old norms. Sentries were posted. No interunit visits were permitted above squad level, and the mail was heavily censored. Within this strictly controlled environment the cadre, removed from the responsibility of conducting combat operations, moved among the three-man cells concentrating on reestablishing comradely relations and operating the informants system. One rallier, a private, whose unit suffered from hardships and heavy combat casualties, described such a situation:

> People were only able to escape at night, and it was very difficult to do it because we were closely watched. . . . It was difficult to cross the VC-controlled area without any food. If the VC laid hands on those who escaped they would send them immediately to a thought reform session.[73]

There is no evidence that the PLA was forced to rely to any great extent on the technique of isolation to regain control of their units during the 1966-1967 period. For those units where it was used the results were remarkable. In a few short weeks they emerged from isolation "unafraid of death" and prepared to resume combat operations.

In sum, whatever the balance in the various PLA units between the leadership of the comradely leader, surveillance within the three-man cells, direct supervision, self-criticism, individual criticism, group criticism, and isolation, there was no disintegration, collapse, or significant loss of military effectiveness within the PLA. The PLA endurance under such hardships and combat stress which would have led to the disintegration of a lesser army is a tribute to PLA organization and leadership.

INDOCTRINATION: CREATING GOOD COMMUNIST SOLDIERS

The final function performed by the cadre was creating "good communist soldiers" through indoctrination. The usefulness in examining the relationship between the organization (the party), the cadre-leader, and the primary group for assessing the impact of indoctrination on a soldier's willingness to fight has been suggested by a number of investigators. Amitai Etzioni, in a review of studies by Janowitz, Shils, Stouffer, and others, divides the issue into three broad areas: first, the organization structure, whether it is a party, a nation, or another entity, and the extent to which it succeeds in internalizing its values and goals within the soldier; second, the primary group solidarity or cohesiveness in influencing the individual soldier's behavior; third, the leader's role in influencing group values.[74]

Similarly, Shils and Janowitz, in their study of the Wehrmacht, demonstrate that when combat soldiers in the German army accepted Nazi officers and noncommissioned officers as leaders, the unit was extremely effective in pursuing the military goals of the Nazi party.[75] Thus the leader binds the combat soldier's primary group with a more authoritative, secondary group which is generally a political entity such as a party or a nation. Shils and Janowitz also point out that Wehrmacht units which surrendered or evaded combat often had a strong sense of loyalty toward the cohesive primary group, but did not fight well because the group's primary leaders had not embraced Nazi values.

What is clear from the studies of Shils, Janowitz, George, and others is that if effective and professional combat units are to be formed, some means must be found to make soldiers fight aggressively for the organization's goals. This is often referred to as "internalization," that is, the individual makes the values and goals of an organization his own.

Measuring the effectiveness of indoctrination on the internalization process within military units is difficult. The prevalence of indoctrination may be demonstrated, but like any socialization process, its effects are hard to assess. Although acknowledging that the PLA mounted an extensive indoctrination effort, Douglas Pike believes the effects were not of major significance in explaining why main force units fought so well. Pike states that an important theme of PLA indoctrination, as noted in a main force indoctrination booklet, was to instill "revolutionary spirit." However, Pike attributes main force fighting skill to factors other than the effects of indoctrination:

> Americans and others often assumed that the NLF army members were fanatics. Because they performed well in combat, it was argued they were highly motivated, which meant dedication to an ideological cause. Thus the search for the essence of this belief. It proved elusive largely because it did not exist. The best of the military units—the Main Force units—were highly effective because they were professionals. . .what impelled them was not ideology so much as professional competence, much like the U.S. Marines or the French Foreign Legionnaire. . . . Their mystique should be attributed chiefly to a unit esprit de corps that stemmed from the consensus that each man in the unit was a superior and vastly experienced professional.[76]

Although Pike does not completely ignore the role of indoctrination, he attributes the strength of the PLA to the careful manner in which it was organized. He cites three major factors, similar to those discussed previously, as instrumental in binding the PLA as an organization: first, "shared social myths," mainly those propagated by the party; second, "leader-led" relationships; and third, indoctrinational efforts directed toward the party cadre, not the individual soldier.[77]

A recent study by Paul Berman agrees that the individual PLA soldier generally did not have "strong political beliefs." Instead, Berman makes the case that major secondary group norms, centered around Vietnamese cultural values and peasant socialization, were "awakened and given motivational force" by the PLA.[78]

In assessing the impact of indoctrination, it is important to note that the process of internalization can assume a variety of patterns within combat units. At one extreme is internalization which emphasizes indoctrination as in the Chinese and Vietnamese models. At the other is a passive process within which formal indoctrination plays almost no significant role. In the American army, for instance, this process requires no formal recognition of American values. Professional combat units are produced without formal indoctrination. In fact, the internalization process is often an unconscious or passive acceptance of the political and social values of the accepted leader who has undergone a stronger socialization process. As discussed earlier, the reasons a soldier accepts the values of his leader are many and varied. Often these values are of lesser importance than the relationship between the leader and soldier. United States Army doctrine holds that there is a high correlation between the successful combat leader and such characteristics as fairness, setting the example, honesty, endurance, knowledge of job and situation, integrity, and so on. Formal indoctrination is not necessary. When the U.S. soldier accepts his leaders and decides to fight willingly under their direction, he also accepts, consciously or unconsciously, the values of the organizational structure which appointed these leaders.[79]

On the other hand, a goal of the PLA leadership was to create a professional communist army through indoctrination and other means which would be able to surpass any other in its "human" capabilities. Alexander George states that this is an essential tactic for any army likely to face a better-equipped enemy.[80] The "man over weapons" idea is central to the techniques used in the PLA for promoting motivation, morale, and group solidarity around party norms. Although this is a Chinese notion, the Vietnamese communists copied it in the development of their army. This is confirmed by G.K. Tanham who traces the Chinese model through the Vietminh to the PLA.[81] The party made a conscious decision to pursue political objectives in all areas, in accordance with the Vietcong axiom of "party control over the guns."[82]

As has been discussed, the internalization of party values by PLA primary groups was accomplished primarily by cadre-leaders who were instrumental in inculcating party goals within the individual fighting man. Where the cadre were able to perform linkage functions effectively, PLA units were highly committed to the combat goals set by the party. A great number of variables determined whether the cadre-leaders could effectively accomplish these functions, and these in turn had their impact on the efficacy of indoctrination.[83]

George cites evidence to show that indoctrination became less effective in the face of military and climatic hardships.[84] For some troops, indoctrination became less meaningful over time, and more of a formality against which they could erect mental barriers and dissimulate. For others, the indoctrination seemed more successful. George describes both cases using categories of "in-groups" and "out-groups." The in-group refers to those troops who "identified with and participated actively in the officially prescribed pattern of group life within the unit." While in-group identification with official goals occurred for very complex reasons it was evident by the individual's belief in the "party line" assimilated through indoctrination. The out-group consisted of soldiers "who remained poorly assimilated, and were more or less alienated from the ongoing, officially sanctioned group life of the unit."[85] Of the sixty-one soldiers interviewed by George, forty-six considered themselves out-groupers while only fourteen considered themselves in-groupers.[86] This gives some measure of the success of the communist indoctrination attempt within the Chinese PLA. It must be noted, however, that the Chinese were subject to influences different from the Vietnamese PLA.

In both cases, whether a soldier was a member of the in-group or out-group, he still fought. While the political officer's goal was to create "positive adherents," he often settled for the more modest objective of a soldier who would fight with his unit.

The Vietnamese PLA attempted to use all of the previously described techniques of motivation and control to produce a "good communist soldier." Ideology became a set of standards, expressed in slogans, against which individual and unit performance could be measured by the cadre. How well the soldier performed "ideologically" determined the degree to which the cadre would focus the morale informants system on the soldier. Communist ideology was not understood in detail by the PLA soldier. His indoctrination consisted of learning general slogans and memorizing the various "good points" of a campaign. The PLA soldiers understood that these slogans represented the general path to be followed to the ultimate "goals." It was not necessary to understand these goals in detail, nor the specific policies designed to gain them. It was enough to belong to the group and trust in the cadre leadership.

The following excerpt from a captured document is representative of the depth of ideological understanding expected of the PLA soldier:

> . . .the political education and ideological leadership must consist mainly of explaining to them (the PLA soldiers) the situation and their tasks, improving their revolutionary nature and the armed forces' tradition, beinging into full play their revolutionary heroism, their ardent patriotism, their hatred for the enemy, and their determination to fight to liberate the South, defend the North, and proceed toward unification of the fatherland and to stand shoulder to shoulder with the Cambodian and Lao people to expel the U.S. imperialists from the Indochinese Peninsula.[87]

Another captured document from the elite Dong Nai Regiment outlines a detailed plan for improving the ideological commitment of soldiers assigned to the regiment. The purpose of this plan, according to its author, was to provide guidance to unit leaders in their efforts to "build an appropriate Revolutionary philosophy of life for the Army (What must we do to lead a meaningful life? For what purpose do we live and fight? Whom do we serve? How should we sacrifice ourselves so that our sacrifice can be worthwhile and glorious. . .)." Excerpts from this document illustrate the content of indoctrination at unit level in the PLA.

> Regarding the indoctrination task, attention should be given to the following points:
>
> Strengthen our Revolutionary standpoint which consists of aggressively fighting the U.S. to the end to liberate the fatherland and the people. . . .
>
> Indoctrinate the strategy of the Party. Consolidate a strong belief in victory. . . .Let everybody thoroughly understand the requirements and missions of his unit.
>
> Indoctrinate them with the traditions of the people, the armed forces, and the units. Enhance their pride in such traditions. . . .
>
> Conduct class indoctrinations; encourage a deep hatred toward the enemy; encourage resentment against the regime of the U.S. aggressors and their henchmen. From these attitudes, we can nourish our unit members' patriotism. . . .
>
> Indoctrinate them with gallantry (be absolutely faithful to the Party and the people). They must not surrender to the enemy or betray the Revolution under any circumstances. At all times they must be determined to fight and not to let the enemy capture us. In case we are captured against our will, we should remember that the Revolutionary soldier is always in the position of the victorious, in the position of attacking the enemy; therefore, we must never give the enemy any information whatever. We must recall the indomitable example of our elders and of our comrades Nguyen Van Troi, Nguyen Van Be, Nguyen Duc Thuan. . . .[88]

If an ideology may be simply defined as a normative view of past history with an assessment of the present along with a plan for the achievement of a utopian future, the content of VC ideology as perceived by the rank-and-file PLA soldier was quite simple and straightforward. The past was almost always referred to in terms of colonialist exploitation by imperialists and their puppets. For example:

> . . .The Americans were like the French before. The French came here because, according to them, the Vietnamese people were stupid and needed their help. Actually they

> came here to rule over the Vietnamese people. Now the Americans aren't much different from the French. . . .[89]

The present was seen in terms of a just and necessary war: "We are fighting in order to liberate the country and its people." Reference to the future was vague, but often included a reference to the achievement of a just society. ". . .this war is being fought to eliminate injustice and exploitation of the poor by the rich."[90]

While the catch slogans and ideological "rules" cited by the interviewees asked about their ideology have a Marxist-Leninist flavor, they were not indicative of any in-depth study or understanding of Marxist-Leninist or Maoist thought. In most cases, the PLA soldier merely had the capability of parroting the party line learned from the cadre.

The ideological underpinning of lower-level cadre was also very shallow. Even those who attended PLA military schools received little education in basic communist ideology. The following excerpt from the interview of a PLA lieutenant, a party member and graduate of a military school, illustrates the limited extent of his formal ideological education:

Question: Did you have to learn about Marxism-Leninism when you join (*sic*) the Party?
Answer: I did but very few. . . . I learned only the rules.
Question: What are the books you were required to read?
Answer: None. But they recommended that you read Marxist books. However, you are free to read or not. . . .
Question: What do you know about Marxism?
Answer: A little bit. Marx taught that he will bring peace and prosperity, a peaceful and equalitarian world. Everyone equal. To each according to his needs from each according to his abilities.
Question: What are the ways to get to that brave new world? Did they teach you that?

Answer: They taught me to believe in Marxism-Leninism. Then there would be a general rise toward Communism. First socialism then communism. What people ought to follow is the materialism of the ideology. What a peasant should do, a city-dweller should do, a worker should do.

Question: Do you think it is Vietnamese?

Answer: This is very difficult to say. As for me I think there are many points that (are) not compatible with Vietnamese society.[91]

How much credit should be given to PLA indoctrination in explaining why the PLA soldier fought so well is an impossible question. While the effects of indoctrination no doubt varied from unit to unit and with individual soldiers, any measurement of the impact of indoctrination must entail an assessment of the total motivation-and-control system implemented by the PLA. Indoctrination was an integral part of the PLA's system. In a very real sense ideology became the medium through which the men, the cadre, and units interacted, providing standards of behavior. Broad ideological goals such as "achieving national independence under the leadership of the Party" and "building a just society" were linked by the cadre to detailed rules covering all aspects of the soldiers' existence.[92] In this way, indoctrination provided a detailed set of standards for the guidance of individual and unit behavior. It also provided a basis for the measurement of individual and unit performance, and was of immense aid to the cadre in the implementation of the morale informants system.

NOTES

1. Philip Selznick, *The Organizational Weapon* (New York: McGraw Hill, 1952), p. 19.
2. Ibid., p. 20.
3. Ibid., p. 18.
4. Frank S. Meyer, *The Moulding of Communists* (New York: Harcourt, Brace, 1949), p. 14.

5. Douglas Pike, *Viet Cong, The Organization and Techniques* (Cambridge, Mass.: The MIT Press, 1966), p. 374.

6. J. J. Zasloff, *The Viet Minh Regroupees* (Santa Monica, Calif.: RAND Corporation, 1966), p. 2.

7. Ibid.

8. Ibid., pp. 3-4.

9. Document no. 72, *Viet Nam Documents and Research Notes*, General Song Hao, "Party Leadership Is the Cause of the Growth and Victories of our Army," American Embassy, Saigon, 1970. This article is the third in a series of three by General Giap, Lt. Gen. Van Tien Dunc, and General Hao. It was serialized by *Quan Doi Nhan Dan* and Radio Hanoi from 24 to 26 December 1969. The author, General Hao, as the top political officer in the PLA, reasserts that man is the "decisive factor in battle" and that the cadre guiding the army as the agents of the party are central to the PLA's struggle to achieve party goals (emphasis in the original).

10. Pike, *Viet Cong*, pp. 373-75.

11. Alexander George, *The Chinese Communist Army in Action* (New York: Columbia University Press, 1967), pp. 63, 86.

12. Michael C. Conley, *The Communist Insurgent Infrastructure in South Vietnam* (Washington, D.C.: Center for Research in Social Systems, American University, 1966), p. 150.

13. David W. Elliot and Mai Elliot, *Part Three: Military Organization and Activities*, RM-5850-ISA/ARPA (Santa Monica, Calif.: RAND Corporation, 1969), pp. 64-65. (Within the demolition platoon of the 514th Battalion the cited passages indicate these policies were the practice.)

14. Pike, *Viet Cong*, p. 231.

15. Elliot, *Military Organization*, pp. 64-65.

16. George, *Chinese Communist Army*, p. 86.

17. Ibid., pp. 88-89.

18. Ibid., p. 89.

19. Ibid., p. 90.

20. Interview K-5, p. 24.

21. A more detailed discussion of indoctrination in the PLA is provided later in this chapter. The purpose here is to show its relation to the morale informants system.

22. This was especially the case during the 1965-1967 time frame when to my knowledge no PLA units defected or surrendered. This is significant as it indicates that the cadre within the PLA had a firm grasp on their roles and were able to bring their units through a period in which the combat power focused upon them underwent tremendous escalation.

23. Document no. 72, *Viet Nam Documents*, p. 4.

24. Ibid., p. 30.

25. Ibid., p. 34.

26. Interview K-11, p. 11.

27. Amitai Etzioni, *Complex Organization* (New York: The Free Press, 1961), pp. 58-59.

28. See Introduction for review of individual soldiers' needs in combat.

29. Interview K-8, p. 25.

30. For a review of a number of such studies, see E.P. Hollander and J.W. Julian, "Contemporary Trends in the Analysis of Leadership Processes," *The Study of Leadership* (West Point, N.Y.: USMA Printing Plant, 1970), pp. 4-4/4-19.

31. William D. Henderson, "An Integrated Concept of Leadership: A Functional Model," in *The Study of Leadership* (West Point, N.Y.: USMA Printing Plant, 1972), pp. 3-4/3-5. Note: This definition of leadership was adapted from Robert Dahl's definition of political power in "The Concept of Power," in Nelson Polsby et al., *Politics and Social Life* (Boston, Mass.: Houghton & Mifflin, 1963), pp. 106-19.

32. Many studies of group dynamics and leadership discuss the sources of leader influence within the group. The particular categories of leader power used here are from J.R.P. French, Jr., and B. Raven, "The Bases of Social Power," *The Study of Leadership* (West Point, N.Y.: USMA Printing Plant, 1970), pp. 7-3/7-17.

33. David W. Elliot and Mai Elliot, *Part Two: Documents of an Elite VC Delta Unit,* RM-5849-ISA/ARPA (Santa Monica, Calif.: RAND Corporation, May 1969), p. 169.

34. Document no. 15, *Viet Nam Documents.*

35. Interview K-14, p. 2.

36. Interview K-15, p. 26.

37. Frank Denton, *Volunteers for the Viet Cong,* RM-5647-ISA/ARPA (Santa Monica, Calif.: RAND Corporation, September 1968), p. vi.

38. Interview K-21, p. 7.

39. Interview K-41, p. 14.

40. Interview K-9, p. 33.

41. Interview K-63, pp. 1, 15.

42. Interview K-19, p. 23.

43. Interview K-15, p. 3.

44. As the war intensified in many areas, the PLA found itself less welcome as its presence was often associated by the villagers with United States and ARVN attacks.

45. Interview K-5, pp. 7, 15.

46. Document no. 72, *Viet Nam Documents* (emphasis in the original).

47. Interview K-5, p. 10.

48. Interview K-9, p. 6.

49. Interview K-20, p. 4.

50. The PLA attempted to minimize this effect whenever possible by withdrawing units suffering from extensive casualties for a thorough re-

building. Such units were usually isolated and remained so until the new cadre had full control.

51. Interview K-12, p. 9.

52. Document no. 72, *Viet Nam Documents*, p. 41 (emphasis in the original).

53. Interview K-5, pp. 8-9.

54. Interview K-12, p. 7.

55. Interview K-22, pp. 7-8.

56. Interview K-4, pp. 5-6.

57. A similar but much more extensive situation was experienced by the Chinese PLA in Korea. Starting in early 1951, the U.S. Eighth Army regained the offensive and inflicted heavy losses and hardships upon the Chinese PLA causing an almost complete breakdown of the "comradely leader" approach to motivation. This forced a rapid Chinese shift to the morale informants system in order to control their army in the face of continued United States pressure. Of course the Chinese PLA in Korea in 1951 was not nearly as sophisticated as the Vietnamese PLA was in 1967 and the crisis for the Vietnamese was not nearly as serious as it was for the Chinese. The significant point, however, is that both armies did turn to the morale informants system with its surveillance and control mechanisms in a difficult situation which could well have led to the disintegration of a less well-organized force. For more details of the Chinese experience see George, *Chinese Communist Army*, pp. 163-96.

58. As stated earlier, the morale informants system has another major purpose in addition to surveillance and control. It complements the resocialization efforts of the cadre-leaders in indoctrination programs aimed at creating not just "good soldiers" but "good communist soldiers." This subject and the associated questions concerning the extent to which party/PLA values were internalized, the problem of dissimulation, and the erosion of commitment under the pressures of combat, are discussed in Chapter 6.

59. Interview K-20, pp. 22, 24.

60. Interview K-10, p. 16.

61. Interview K-3, p. 13.

62. Interview K-7, p. 9.

63. Interview K-15, p. 23.

64. Interview K-4, pp. 7-10.

65. Konrad Kellen, *A Study of Motivation and Morale*, RM-6131-I-ISA/ARPA (Santa Monica, Calif.: RAND Corporation, September 1970), p. 30.

66. Interview K-5, pp. 28-30.

67. Kellen, *Study of Motivation*, p. 32.

68. Ibid., p. 33.

69. Ibid., p. 23.

70. Interview K-28, p. 21.

71. This comment does not pertain to training units supporting PLA main force units where the standard practice, as it is in most armies, was to separate the recruit from past relationships.

72. Interview K-20, p. 24.

73. Interview K-2, p. 4.

74. Amitai Etzioni, *Complex Organizations* (New York: The Free Press, 1961), pp. 58-59.

75. E. A. Shils and M. Janowitz, "Cohesion and Disintegration in the Wehrmacht in World War II," *Public Opinion Quarterly* 12 (2), 1948, pp. 286-88.

76. Douglas Pike, *Viet Cong, The Organization and Techniques* (Cambridge, Mass.: The MIT Press, 1966), p. 377.

77. Ibid., p. 377.

78. Paul Berman, "The Liberation Armed Forces of the NLF, Compliance and Cohesion in a Revolutionary Army" (Ph.D. dissertation, MIT, 1970).

79. Etzioni, *Complex Organization*, p. 58.

80. Alexander George, *Chinese Communist Army*, p. VI.

81. George K. Tanham, *Communist Revolutionary Warfare* (New York: Praeger, 1967), pp. 33-35, 136-70.

82. David W. Elliot and Mai Elliot, *Part Two: Documents of an Elite VC Delta Unit*, p. 1.

83. Any number of factors can interfere with this linkage process. Loss of "accepted" leader status by defeat in combat, prolonged deprivation of food, inadequate medical care, depletion of personnel strength, and other combat situations can attenuate the effectiveness of indoctrination efforts.

84. George, *Chinese Communist Army*, pp. 130-32.

85. Ibid., p. 81.

86. Ibid., p. 81-82. This sample was drawn from Chinese soldiers in Korea in 1951.

87. Document no. 94, *Viet Nam Documents*, "Ideological Problems Within the People's Liberation Armed Forces," p. 20.

88. Document no. 46, *Viet Nam Documents*, "On Political and Ideological Indoctrination Against Desertion and Surrender," pp. 4-5.

89. Interview K-1, p. 28.

90. Interview K-5, p. 49.

91. Michael C. Conley, *Communist Insurgent Infrastructure*, pp. 302-03.

92. As described earlier, the degree to which the PLA soldier "internalized" these rules was heavily dependent upon his immediate cadre and the degree to which the cadre was the accepted leader of the group.

6 The Effects of Hardship and Combat

Within the PLA from 1965 to 1967, it appears that stress and hardship of combat were major factors in eroding cohesion in various PLA units. In those units where the skill of the cadre in countering these factors was not successful, severe motivation and control problems arose.[1] Even PLA soldiers who at one time could be described as thoroughly dedicated to the PLA cause wavered and deserted in the face of unmitigated hardship and stress.[2]

The following excerpts from captured diaries of PLA soldiers sharply contrast original enthusiasm with bitter realism after a period of substantial combat and hardship. One diary belonging to Mai Van Hung provides a record of his steady decline in morale as he experienced food shortages, disease, homesickness, and fear of death.

> 24 August 1965: I am leaving the camp tonight. My heart is filled with love for my homeland. I pledge to achieve victory before returning to my homeland.
>
> 4 September 1965: . . .We hold no hope of life. No words can express the hardships of our lives. I feel pessimistic and downhearted. Can anyone understand my inner feelings?
>
> 1 January 1966: No vegetables and meat can be found here. We have nothing for food except salt, salted shrimp paste, and dried fish. How unbearable life is. . . .In what way does the war benefit them? (The reference to "them"

is an apparent criticism of his immediate superiors or of the VC leadership.)

18 July 1966: . . .The horrible disease deprives me of sleep and appetite. I now envision so many horrid scenes. My heart is filled with bitterness. I wish I could leave the hospital, but the frightful paralysis keeps tormenting me. . . .

28 July 1966: . . .My sickness does not seem to improve. Who can understand me? What would they take me for? ("They" again refers to his superiors.)

29 July 1966 (last entry): Time keeps going its steady course. Life remains unchanged.[3]

Excerpts from another diary also illustrate the gradual change in morale from a poetic exaltation in the pursuit of PLA goals to disenchanted realism.

3 July 1967: . . .Youth must be like the flame tree flowers and stand firm in their determination to face the rigors of war.

11 August 1967: . . .No matter what rigors are imposed on him, he remains unyielding in his opposition to the Americans. . .he still carries on the fight without backing up as long as there is one single American still alive.

21 September 1967: . . .Feel very hungry. All my limbs become numb; have ringing in my ears; feel giddiness in my head.

4 October 1967 (disjointed comments—last entry): Weak ideology, fear of hardships, poor observance of regulations. Guard. Hunger. Wish to live in low land area. Poor execution of orders. Poor sense of organization.[4]

The problems faced by the cadre attempting to maintain a high degree of motivation and control in the face of sustained hardship and combat stress are well summarized by the following quote from a PLA fighter.

. . .when the fighters are hungry they don't want to fight no matter how many lectures they have heard. When the fighters are hungry, they refuse to carry out missions and go on operations. What could the cadres do? Morale is important, but materiel is also important.[5]

A soldier's ability to resist an overwhelming concern for his personal security and to carry out an effective combat role largely rests on his unit's capability to satisfy basic security and social needs. When his unit is unable to meet such needs as food, medical care, safety, and comradeship, or when it is threatened by attrition, it becomes much less effective in controlling the behavior of soldiers. This is reflected in the following comments by a rallier:

Question: How would you describe the morale in your unit?

Answer: When I first joined my unit, the morale was relatively good. But day by day there were more and more bombs and shells, the casualties kept increasing, and defeats followed one after another. Therefore, day by day, the fighters became more and more discouraged. They went on fighting only because they were forced to do so. There was a regroupee cadre who had operated in Tuy Phuoc District and who, after having rallied to the GVN, called upon us to rally through leaflets dropped by aircraft. Many fighters, after having read these leaflets, wanted to rally but couldn't do it because they were closely watched.

Question: How many casualties has the 21st Regiment suffered up to now?

Answer: It was said by the Regimental commander himself that after having engaged in battle with the Americans at the beginning of the year in Hoai Nhan District, it was found that the number of killed, wounded, and those who got lost or were captured, amounted to 700 people in all. Also in July when the Americans conducted a sweep operation in Song Con mountain, the 9th Communication Company actually lost 25 people composed of a political officer, an assistant company commander, two signal leaders, two platoon leaders and fighters. . .(Paused). . .In February, there were also more than 25 people who deserted. Therefore, there are only 60 or 70 men left in my company, which originally totaled 120 men.

Question: Which attacks caused your division the most damage and losses?

Answer: First, it was the artillery, second, the helicopters, and third, the bombers. By now in the front they don't fear the Morane (L-19 unarmed spotter aircraft) any more. They are afraid of a kind of helicopter they used to call guitar-handle-helicopter or pig-basket, or red nose helicopter because many had noses painted red. . . .It was very difficult for us to hide ourselves from that kind of helicopter because they flew very low and thereby, were able to spot targets very easily . . .that kind of helicopter strafed fiercely with their .50 caliber machine-guns (actually 7.62 mm.). They once killed five guerrillas who had already taken cover in commo trenches. . . .[6]

In response to a question concerned with the stresses of combat, another rallier stated:

> I had a 99 percent chance of being killed. That's why I decided to quit. . . .I was afraid only of the B-52s. They dropped bombs as if they were handfuls of sand at random. It was impossible to dodge them all. . . .The greatest losses occurred when the battalion was bombed by the B52s. We lost 30 dead and wounded (in one raid). . . .[7]

In response to a similar question, another rallier replied:

> We didn't talk about surviving the war because we knew that we might get killed any time. None of us believe that he might get out of the war alive.[8]

The hardships of guerrilla life also seemed to have a deleterious effect on the soldier's motivation to carry out PLA missions. This appears to have been directly linked to the strength of primary group norms (including ideological ones) over the soldier. At times of rapid personnel turnover, prolonged deprivation of food, and inadequate medical care the primary group had a decreasing ability to offer esteem and other associated needs, and the norms of the group (ideological norms included) became less of a determinant of the soldier's behavior. A disintegrative trend based upon the individual's preoccupation with physical survival developed as the primary group's cohesive glue weakened.

When questioned whether the men in his unit (primary group) trusted each other, which is a basic requisite for success in combat, a rallier replied:

> We couldn't trust each other much because there had been changes in the unit's composition. Some members got killed and were replaced by the new ones, some were sent away for other missions, others were transferred to other companies in the same unit, and new ones were transferred to my company. Some cadres were sent to other companies in the same unit. These transfers back and forth weakened the intimacy among the men in my company.[9]

The same rallier describes how hardships fostered "incorrect thoughts" among his "close friends" (that is, the primary group), which in turn fostered deviant norms as far as the PLA was concerned.

> The worries and discontent toward the VC cadres were privately discussed only among close friends. We could not talk to our cadres or comrades who were our friends, because the political officer judged us on what we said. Our worries mostly concerned our present life, like our starvation, our lack of medicine, our increasing casualties in an intensive war, and our bad health that cut down our strength. In general, we were afraid for our fate. We did not know what we fought for. Who would benefit if we died? As most of my comrades. . .were single, they would be survived by nobody since their old parents will die some day. They found their fighting without purpose and the future victory no use to anybody.[10]

The skill of the cadre was crucial in mitigating the impact of combat and the hardships of guerrilla life, and if the cadre were unskilled or inexperienced, the PLA suffered.

Question: Describe the cadres in your company. What kind of persons were they? How much experience have they had?
Answer: They were newly appointed cadres. They just got out of Officers' Candidate School. They didn't have much experience.
Question: Did you have confidence in them?
Answer: Not much.
Question: Why?
Answer: Because they were still very young—between 22 and 30—and also because they just got out of school. Some of them had little education—two years in school.
Question: Who was in charge of political education and morale problems in your unit?
Answer: The Political Officer.
Question: What did he do especially well?
Answer: I think his work was just so and so. . . . When he was scheduled to give a lecture, he gave a lecture; otherwise, he stretched out and went to sleep like the other fighters. . . .When the men were sick they refused to go on operations, they remained in bed. When the men were discouraged, they feigned sickness and made excuses for themselves so they could stay behind. They might be demoted and expelled from the Labor Party, but they stayed in bed all the same.[11]

Another rallier, generally unimpressed with his cadre, describes why:

Question: What did the political cadre do? What were his responsibilities?

Answer: The political cadre's main mission was to excite the fighters' courage and enthusiasm before they were sent on combat operations. He aimed at getting the morale fighting spirit of the fighters high in any circumstance. To attain the goal, he used to conceal the Front losses from the fighters and exaggerate the ones suffered by the Allied forces. But most of the people in my unit have come to know that the political cadre's statements were indeed far from truthful.

Question: What did you think of the political cadres in general?

Answer: I showed some regard for the battalion political officer, but not for the lower political cadres. In my unit two company political officers have been replaced for having made unsound statements.

Question: Could you give me an example of what you felt was unsound?

Answer: For example, they said, "We inflicted on the GVN troops a loss in men amounting to two full companies, and we captured two carbines." Those cadres didn't know how to lie.[12]

These statements support the findings that if an army such as the PLA has unskilled, inexperienced, unenthusiastic, graceless cadre, it can do little to influence the attitudes of the soldiers in adversity. On the other hand, if an army has well-liked, respected, skillful, experienced men serving in leadership positions, it will have some success in mitigating the stresses and hardships of war. This is well illustrated by the responses of this reluctant rallier:

Question: What were the greatest hardships you and your unit had to suffer?

Answer: I and people in my unit were afraid most of continuous operations during which one had to carry on his back a load of some thirty kilos on a long march. I think as long as I live I will never forget those hardships. Once I had an opportunity to visit my family, everyone cried in looking at me because my body was completely losing its shape. I was in really poor condition after a short time in the Front forces. Then I was heartbroken upon learning that my family had been living in misery because of my absence. Nevertheless, my mother found a way to give me some money when I left for my unit. I was in tears when I took the money because I knew that it was borrowed from a neighbor.

Question: Who was in charge of political education and morale problems in your unit?

Answer: My unit had an excellent political cadre. He was very skillful in convincing people, particularly those who worried about their families (the respondent was particularly worried about his family). He used to get in contact with soldiers in private to advise and comfort them. Everyone liked him and followed his advice.

Question: Do you think his mission was an important one? Explain.

Answer: I think his mission was the most important one. Thanks to him the soldiers could keep their fighting spirit at a fairly high level. . .his behavior, attitudes, and performance always remained the same.

The same rallier also saw other cadre in a similar light.

Question: What other cadres helped mobilize the spirit and watch over the correct behavior of the men?

Answer: All the assistant platoon leaders were people who helped the political cadre mobilize the spirit and watch over the correct behavior of the men. They received reports from the squad leaders about the soldiers.[13]

These excerpts show that the PLA was not completely successful in building a professional army composed of "good communists." The success of their control system and indoctrination efforts varied from unit to unit and within units over time. This may be illustrated by applying George's categories of in-groups (beets) and out-groups (radishes) to the PLA.

A cadre (rallier): "I knew that about two-thirds of my men freely expressed their discontent and depression."

Discussing the political officer's success in indoctrinating his unit, a fighter (rallier) states:

> Impartially speaking, the Company Political Officer could influence a certain number of troops who always trusted and liked him. Those who were discontented with him or had a neutral attitude never trusted him, but just complied with his orders for the moment. The Political Officer was successful, somewhat, in raising the morale and fighting spirit of those he could influence. He used these men as an example for the discontented and undecided men in his unit. In short, my Political Officer was a clever leader. He usually talked privately to each individual and psychologically said, "We must show the people in South Vietnam that we have lived under socialism in North Vietnam, etc." However, he could actually control only half of the people in my unit. . . .[14]

A company commander (prisoner) stated: "About two-thirds of the men were high-spirited and eager for combat duty. Others were less enthusiastic because they didn't like the hardships and danger."

It would, of course, be extremely difficult to state to what degree the PLA was successful in creating "good communist soldiers" through their indoctrination program. Table 2 shows that out of the thirty-one interviews examined for data (in-group versus out-group), thirteen interviewees considered themselves outgroupers.[15] Not unexpectedly, the ingroupers were mostly prisoners (two ralliers were also ingroupers) while the outgroupers were all ralliers. However, it is not possible to generalize about the degree to which outgroupers were evident within the PLA from the figures in Table 2. The range that has been suggested goes from one-half of one unit to zero in another unit. With the surveillance and control mechanisms

Table 2

The Effects of Hardship and Stress on PLA Indoctrination and Cohesion
(numbers reflect interview numbers)

	POW		*RALLIER*	
	Ingrouper (Positive Attitude)	*Outgrouper (Negative Attitude)*	*Ingrouper (Positive Attitude)*	*Outgrouper (Negative Attitude)*
Interviews Reflecting Stress of Combat	None	None	None	2, 4, 6, 7, 10, 42 (Total - 6)
Interviews Reflecting No Stress of Combat	3, 5, 8, 9, 11, 17, 19, 21, 26, 30, 51, 56, 63 (Total-13)	None	None	1, 27, 28, 36, 37, 39, 46, 60 (Total - 8)
Interviews Reflecting Hardship	None	None	None	1, 2, 4, 6, 7, 10, 27, 28, 36, 39, 40, 42, 44, 46, 47 (Total - 15)
Interviews Reflecting No Hardship	3, 5, 6, 7, 8, 9, 11, 12, 19, 21, 26, 30, 51, 56, 63 (Total-15)	None	None	20, 27, 37, 60 (Total - 4)
Totals (interviews only counted once)	16	0	0	18

Terms:

POW: Prisoner of War
Rallier: PLA defector to RVN forces
Ingrouper: Reflected a positive attitude toward the party line and participated actively in the officially prescribed pattern of group life within the unit.
Outgrouper: Reflected negative attitudes, was poorly assimilated, and remained apart from the officially sanctioned group life of the unit.

utilized by the PLA to prevent defection, it is unlikely that an accurate accounting of the number of outgroupers in the PLA can be determined.

Table 2, in the first column, reveals that the ingroupers generally experienced little stress of combat or hardship, while many ralliers had been exposed to intense stress in combat and had endured prolonged hardship. Moreover, all POWs or ingroupers reflected a significant degree of indoctrination while the ralliers or outgroupers generally did not adhere to the party line.

Table 2 suggests that adherence to the party line in carrying out duties was greatly dependent upon stress in combat and prolonged hardship. The greater the stress and hardship the less effect indoctrination had on the behavior of the PLA soldier. To the extent that the small group was able to cope with stress and hardship and meet the PLA soldier's needs group cohesion was maintained. In almost all cases, those PLA soldiers who defected cited a significant deterioration in unit cohesion. There are only four cases which deviate from this proposition, and they can be explained through special circumstances.[16]

In conclusion, indoctrination alone is not the reason why the PLA endured as it did. Other factors of more significance must be relied upon in conjunction with indoctrination if the remarkable cohesion and combat effectiveness of the PLA is to be explained. Indoctrination appeared to have a significant effect on the willingness of the individual PLA soldier only to the extent that its content (that is, symbols, slogans, etc.) became the criteria around which primary group norms were formed. In other words, no communist ideologues or true believers were found among the interviewees. Those who did recite the party line (superficially) did so primarily because to do so was a norm of their group.

Even though the PLA indoctrination and reeducation program was not always as successful as the party might have hoped, what is important is that the PLA soldier, whether he was a member of the in-group or the out-group, fought for PLA goals. This is supported by the fact that the PLA survived as an organization. There was no widespread disintegration and there was no mass defection of platoon-size or higher units. This was probably due to other aspects of the PLA motivation and control system, such as surveillance and *kiem thao*, directed toward the maintenance of the small group. Perhaps the most important function of indoctrination was the creation of a pool (that is, the in-group) from which reliable replacement cadre could be drawn to fill more responsible positions.

NOTES

1. For example, the party used a measurement similar to Alexander George's categories of "in-groups" (beet-red throughout), and "out-groups" (radish-red only on the outside). When more than 20 percent of a PLA unit was judged "unenthusiastic fighters," the unit was usually removed from combat, taken to an isolated sector, and put through a retraining and reeducation process.

2. Interview after interview made clear that attack from the air (B-52s and helicopters) and artillery appeared to be the most direct cause of combat stress (anxiety caused by danger and fear), while constant U.S./ARVN ground operations were the direct cause of much PLA hardship (lack of food, sleep, medicine, etc.). Ground operations indirectly contributed to PLA combat stress by making PLA units move continually, thus exposing them to air and artillery attack.

3. Document no. 1, *Viet Nam Documents and Research Notes,* "Diary of an Infiltrator," American Embassy, Saigon, 1967, pp. 3-5.

4. Document no. 10, *Viet Nam Documents,* "From Poetry to Reality," American Embassy, Saigon, 1967, pp. 3-11.

5. Interview K-1, RAND Corporation, Santa Monica, Calif., p. 8.

6. Interview K-2, pp. 2-3.

7. Interview K-20, p. 36.

8. Interview K-1, p. 26.

9. Ibid., p. 16.

10. Ibid., pp. 11-12. As explained by Hickey, "Death is no real departure from the family." What is extremely important is that the ancestral lineage remains unbroken, and this is what worried these single, unmarried soldiers who had not produced a son. Their untimely death in the PLA could lead to a break in their immortal family lineage. See G. C. Hickey, *Village in Vietnam* (New Haven, Conn.: Yale University Press, 1966).

11. Interview K-1, p. 8.

12. Interview K-13, p. 5.

13. Interview K-4, pp. 4-5.

14. Interview K-13, pp. 4-5.

15. See Table 2 for definitions of "ingrouper" and "outgrouper."

16. Interviews K-20, K-27, K-37, and K-60. For example, K-20 was a company commander who had been relieved for gross misjudgment in combat, and saw little future for himself within the PLA. K-37, although he could have been content to stay, rallied because he also saw no future in the PLA. He stated, "I couldn't be promoted because they maintained that I was a bourgeoise element. . . . I was dissatisfied because I wasn't admitted to the Party. . .so I rallied."

7

Summary and Conclusions

Modern warfare requires that the successful army rely upon some type of internalized or normative control scheme for the motivation of its soldiers. The dispersion, confusion, chance, danger, and tactics which characterize today's battlefields make obsolete past techniques of control which were based upon drill, rote, and continuous supervision. As a result, the focus of decision and control on today's battlefield has shifted downward toward the squad and platoon levels. This is especially true in guerrilla warfare where it is constantly necessary for the small unit to improvise as it fights and to adapt military objectives to an ever-changing situation. The need for an internalized sense of discipline is a requisite if each individual is to contribute his maximum toward meeting an army's objectives.

The leadership of the PLA recognized these requirements and created a professional army surpassing any other in its "human" capabilities. It was this army which fought in Vietnam during the 1965 to 1967 period investigated by this study.

In answering the central question—why the PLA soldier fought so well in the face of overwhelming firepower and hardships—it was shown that the PLA soldier's values, attitudes, and behavior were shaped by three strong forces: (1) party organization and ideology; (2) the cadre-leaders; and (3) a primary group of soldiers with strong and homogeneous values. The role of the organization was to determine objectives and prescribe values. The role of the primary group was to achieve these organizational objectives in accordance with party values. The role of the cadre-leader was to link the organization and its objectives to the primary group in such a manner that the primary group's values and norms were the same as those of the organization. In the interaction among these three forces the role of the cadre-leader was a key one. As a dedicated party member the cadre-leader, through his role of comradely leader and as operator of the morale informants system, transmitted party/PLA goals to the fighters and led them in their attempts to accomplish these objectives. For their part, the fighters, as members of a primary group with a powerful influence on their individual behavior, fought for PLA objectives under the guidance of the cadre-leader who dominated the group as the "accepted" combat leader and also as the manipulator of the surveillance and criticism mechanisms.

Regarding the organization of the PLA, there are four points of particular importance to the overall function of transmitting party goals to the PLA. First, in terms of internal organization and practices, it was determined that there was little difference between main force units from North Vietnam and those originating in South Vietnam. Second, the time-honored principle of "party control of the guns" was reaffirmed. The descriptions of party control over the PLA from the top echelons in Hanoi down to the basic three-man military cell revealed the important principle of centralized control over policy but decentralized control over the adaptation of policy to local conditions. As for the party functions of PLA cadre, the political and military cadre both were shown as loyal party members, with the military cadre as being more of a technician supporting the furthering of party goals. Finally, the 3-by-3 organization of the PLA was shown to be an efficient means of insuring continuous organizational control down to the very roots of the PLA at the military cell level.

Regarding the primary group as the "cutting instrument" wielded by the cadre in their pursuit of PLA objectives, the central role of the three-man military cell was shown to result from: (1) its function as a "buddy group" capable of satisfying basic needs; (2) its domination by the cadre-leader which insured that primary group norms would be congruent with organizational objectives; (3) its utility as a basic mechanism for surveillance and as a source of reports on individual behavior; and (4) its suitability as a unit capable of engaging in modern guerrilla warfare.

The capacity of the primary group in the form of the three-man military cell to have such a pervasive influence over the behavior of its members was found to be, in part, dependent upon secondary group characteristics such as a common and unique Vietnamese race and history. Commitment to broad secondary group symbols and values such as Vietnamese independence and nationalism also contributed to primary group cohesion. Vietnamese cultural characteristics such as fatalism, respect for elders, "face," and the Vietnamese view of death were also seen as factors which contributed to PLA control over the three-man military cell.

The cadre-leader, by linking the primary group to party/PLA goals, was seen as crucial to the accomplishment of the overall success of party/PLA objectives in South Vietnam. In order to succeed in his three roles of party agent, primary group leader, and chief of surveillance, the cadre-leader had to successfully: (1) transmit party goals to the three-man cell; (2) be dominant within the three-man cell and above; (3) conduct surveillance of individual and group performance to detect deviance and insure compliance with party/PLA goals; and (4) indoctrinate "good communist soldiers." To accomplish these functions, the cadre-leader drew upon the bases of power described and exercised certain control devices such as criticism, surveillance, and isolation.

It cannot be concluded that the PLA was completely successful in creating an army comprised of "good communist soldiers." However, the cadre did motivate and control the "human element" to an unprecedented extent, allowing the PLA to endure the disintegrative effects of sustained combat and hardship during the 1965-1967 period, which resulted in the ultimate control of the battlefield.

Appendix A[1]

THE INTERVIEWS

Between 1964 and 1968, the RAND Corporation interviewed more than 2,400 Vietnamese who were familiar with the PLA and related subjects.[2] In all, there were approximately twenty separate interview projects. One of these, the K-Series, was used as the primary source of this study. These interviews sought to provide information on the PLA organization and why its forces fought so effectively.[3] The following provides additional background on the K-Series and other interview projects conducted between 1965 and 1968.

SCOPE AND LIMITATIONS

It is important that the extremely difficult conditions under which the interviews were produced as well as their limitations and strengths be recognized. A significant limitation is that the interviews do not represent a random sample. Several attempts were made to develop methods for obtaining random samples but these were unsuccessful. As a result, statistical methods of analysis would be risky. As one RAND analyst observed:

> It does not appear feasible to draw any useful objective conclusions from the available data through the use of statistical inference. Furthermore, I can see no possible change in

> the data collection methods which would make such conclusions possible.[4]

The unsuitability of the interviews for statistical analysis does not preclude their use for other methods of analysis. Another RAND analyst stated:

> We are, of course, conscious that any interview program, especially under the conditions prevailing in Vietnam, is bound to be imperfect and that the results will contain a variety of distortions. We believe, however, that the internal consistency of the answers, and the fact that many of the data tie in with information obtained from other sources, indicate that much of the collected information is, on the whole, credible and usable for purposes of analysis.[5]

Another possible limitation is that some interviews are not complete transcripts of questions and answers. Although most interviews were taped, much insignificant material was omitted when the interviews were translated and written up. For example, if a respondent did not completely understand the question, the detailed explanation that followed was omitted. Even though the transcripts were often modified in this manner, diligent effort was expended to preserve the flavor of the interviews. Every attempt was made to insure that the phrases and language used were those of the respondent. Because the process of translating, editing, and writing was susceptible to the introduction of bias, internal checks were conducted.[6]

Another type of bias, which was more difficult to control, was the result of the different personality types and the widely varying opinions held among the respondents. Some respondents were evangelical and attempted to convert the interviewer to their point of view while others told the interviewer what they thought he wanted to hear. When an interviewer thought the respondent might not be answering accurately he would move to another question and later ask similar questions to test the respondent's consistency. Other data was often available for comparison with the respondent's statements. Moreover, the interviewers very quickly became adept at discerning inaccurate responses and often included an assessment of the respondent's accuracy and sincerity at the conclusion of the interview.

Finally, other limitations were imposed by the physical conditions under which the interviews were conducted. The necessity for haste, difficult field conditions, interruptions, the frequent presence of a third party, and the need for improvisation often caused shortcomings in the interviews. Because the interviewers were civilians operating in a war zone, their priority for transportation, quarters, communications, and supplies was often low. Many interviews were edited and typed by candlelight late at night with little relief from the heat and other elements.

In spite of these limitations, however, the interviews should be treated as historical documents and viewed in the context of the situation from which they were extracted. While each interview presents a fragmented and partial view of motivation and control within the PLA, the overall knowledge one gains from the K-Series interviews is why the soldiers at the bottom of the PLA pyramid, who fought the battles, endured the hardship, and survived the combat, remained so tightly bound to the PLA.

THE INTERVIEWERS

The quality of the interviews was heavily dependent upon the skill and integrity of the interviewers; according to W. Phillips Davison, the RAND Corporation was well satisfied on all counts. During the phase in which the K-Series was produced, RAND hired approximately thirty-six interviewers. At any one time the number on the job was usually between ten and fifteen. Several were released when it became apparent they lacked the necessary skills. Others left to take higher-paying and more skilled jobs. While pay varied with age and experience, their salaries were comparable with those received by top ranks in the Vietnamese civil service. As a result, many competent and experienced social scientists and others with professional backgrounds were attracted to the project, and some remained for several years.

All interviewers were Vietnamese civilians and as far as could be determined had no connections with the RVN, the ARVN, the police, or the PLA. Background records of the interviewers are not complete. The information available is displayed in Table 3. Their characteristics are diverse. Three fought with the Vietminh against the French. Twenty-two were born in the North and migrated South. Several were Catholic or Protestant and nine were women.

The most significant characteristic held in common was that none were communists, although this did not mean that they held similar political opinions. As mentioned earlier, three had fought with the Vietminh and it is probable others had also. Several were apolitical, occupied primarily with personal and professional interests. Many who expressed political opinions tended to be very critical of the South Vietnamese government and had been in jail. A former interviewer described the political orientation of the group as follows:

> They all had different political viewpoints and backgrounds, but it can be said that they shared the following common characteristics. They were disenchanted with the GVN (Government of Vietnam), they were sympathetic to some of the goals of the Front (social justice, distribution of land to the peasants, a fairer distribution of wealth among the population, for example), they sympathized with the Front members and understood why they had chosen to join the Front. The interviewers could be called reformists, because they wanted to bring about social and political reforms, but without bloodshed. They all would prefer to see South Vietnam remain non-communist, but some of them wondered whether we should pay any price to keep it so (whether we should continue the killing and destruction to drive the communists out). Perhaps it can be said that they all wished to see their country become one day prosperous and democratic like Japan.[7]

Table 3

Attributes of Interviewers, 1965-1968

ATTRIBUTE	*NUMBER* *n = 36*
Sex	
Male	27
Female	9
Date of Birth	
Before 1920	3
1920-1929	8
1930-1939	16
1940-1949	7
No data	2
Place of Birth	
North Vietnam	22
South Vietnam	5
Central Vietnam	6
Cambodia	1
No data	2
Education	
Completed secondary school	5
Completed university (including graduate school)	16
No data	15
Previous Occupation	
Secretarial or clerical	3
Professional	14
Business	3
Government service	2
South Vietnamese Army	6
Student	3
No data	5
Religious Preference	
Vietnamese traditional (mainly Buddhist)	14
Roman Catholic	7
Protestant	2
No data	13

SOURCE: W. Phillips Davison, *User's Guide to the Rand Interviews in Vietnam* (Santa Monica, Calif.: RAND Corporation, 1972), p. 16.

Because they had no previous experience with interviewing, the interviewers were trained prior to assuming responsibility for conducting interviews on their own. They served first as interpreters for American interviewers, translating questions and answers in English and Vietnamese. In this manner they quickly became familiar with effective interview techniques. Eventually, the new interviewer would be entrusted with conducting several interviews and writing up the reports from tape recordings and notes. The first interviews were carefully reviewed in detail by an American team leader. Ways in which the interview could have been improved were discussed—for example, where the interviewer could have probed more, any inconsistencies left unexplained, promising avenues which could have been developed more fully, and so on. Once the interviewer was considered proficient he was sent into the field subject to further routine checking and editing.

An important question, of course, was whether the interviewers' biases affected their written reports. Because the interviewers were given wide latitude in conducting their interviews, the possibility of bias cannot be excluded. However, the RAND field and project directors stated with confidence that "virtually all" of the interviewers tried to be objective and present information accurately.

A natural tendency might have been to "give the Americans what they want to hear" but the fact that the interviewers were Vietnamese nationalists militated against that tendency. In this regard a field director stated:

> There was amusement and even perverse pleasure in finding out that the Americans were failing so miserably in their efforts to cope with the NLF (composed of Vietnamese), especially when the Americans had raised so many criticisms of the Vietnamese in general so often. Therefore, it cannot be said that the interviewers, being non-communists, deliberately pictured the Front in the worst light or made counter-insurgency efforts look more effective than they really were.[8]

There were no cases of intentional distortion or inaccuracy reported by any of the field directors or team leaders. They did recognize that the potential for unintentional bias was present and attempted to minimize the opportunity for it to affect the interviews.

THE LOGISTICS OF INTERVIEWING

In all, interviews were conducted in more than forty different locations throughout South Vietnam. All parts of the country were represented, although a disproportionately large number of interviews were conducted in Saigon, My Tho, Bien Hoa, Danang, Can Tho, Bac Lieu, and Hue. Logistics presented a continual problem and influenced which POWs and ralliers were interviewed. Even though the project was U.S.-sponsored, no transportation means (aircraft or vehicles) were assigned to the project on a steady basis. This required that separate arrangements be made for every trip. Scheduled flights were sporadic and often involved long delays. Once at their destination, interviewers' problems in finding local transportation, accommodations, and suitable food were endless. Trips out of Saigon were usually for five or six days and at the end of that period the various teams often returned exhausted and suffering from the effects of unsafe food and unhygienic conditions in the countryside. In spite of these conditions, morale remained high, even in dangerous unsecured areas, because of a mutually supporting spirit which was characteristic within the various teams.

THE SELECTION OF RESPONDENTS

As has been noted, respondents were not selected by a random process. Given the chaotic conditions in Vietnam during this period, especially in the administration of POWs and ralliers, any sort of random selection process would have been impossible given the resources reasonably available. Even if the random selection of respondents was feasible, the usefulness of such a procedure was questioned. The purpose of the interviews was to determine the nature of the PLA, not to generalize about the respondents. Had some attempt been made to randomly select respondents, much time, according to Davison, would have been wasted on "the typical undesirable respondent."[9] This was a respondent who could say very little about his experiences in the PLA or with the front because he was a young man, a farmer or a fisherman, from a remote area, who had been drafted by the PLA, and had defected or been captured after several weeks. Instead, respondents who were knowledgeable about the PLA and other front activities were sought. Those who had spent years in the PLA were most informative, often in an open and cooperative manner, whether they were prisoners or defectors.

Table 4

Summary of Respondent Data

Number of Interviews	53
Cadre Respondents	25
Soldier Respondents	28
Respondents from South Vietnam	26
Respondents from North Vietnam	27
Ralliers	23
POWs	30
Party Members	21

Copies of the complete K-Series interviews may be obtained from the National Technical Information Service, Department of Commerce, Springfield, Va. 22151.

Locating such respondents was difficult work. There were no comprehensive or reliable lists of POWs or ralliers available. This was generally true at all levels, nationally as well as at specific camps for POWs and ralliers. As a result, respondents were initially available on a potluck basis. Later, by 1965, when knowledge of the interview project had spread within South Vietnam, many promising leads on respondents were given to the interviewers. Still, much diligent groundwork was required before interviews could actually be conducted as the following quote from a team leader indicates.

> There are rumors of 5,000 prisoners held on Con Son Island. Checking with police advisors confirms them. We obtain a list of 28 prisoners to talk with, but when we arrive on the island we discover that "nobody ever heard of them." We plow through 3,500 dossiers, over 40 percent of which have *tinh nghi* (suspect entered in them making them unusable). We screen 350 subjects, talk at great length with 80, interview 35, and obtain 29 usable interviews.[10]

CONDUCTING THE INTERVIEWS

Once a respondent had been selected, there was usually a problem of where to conduct the interview. Although desirable, privacy was very difficult to achieve. Even in police stations and prisons where separate rooms were sometimes available, interruptions were constant as Vietnamese officials entered and left on errands unrelated to the interviews. When separate rooms were not available, a corner of a larger room was often used; however, the noise and other activities were often distracting. Team leaders estimated the percentage of interviews conducted where privacy was "assured" at 55 percent, "good but not assured" at 23 percent, and "poor" at 22 percent.

Establishing rapport with the respondent was a definite part of the interview. During introductions the respondents were always told that their participation would "neither help nor hurt" them and the interviews were generally described as being in support of a broad research project on PLA and front activities. A Vietnamese interviewer described the introductory portion of the interview as follows:

> Most subjects were curious about who the interviewers were, who they worked for, what they were going to do with the information received, and most important if the information would be used against them later. To reassure subjects, interviewers made it clear that their names were of no importance. . .and would not be divulged. At the same time, they could not expect any reward. The interviewer would be very grateful to them for their cooperation, but if they decided to remain uncommunicative, the interviewer would understand and there would be no complaint

> against them. . . .Some interviewers . . . introduced themselves as researchers for a private U.S. organization. . . .[11]

An important part of building rapport was the technique of offering a package of cigarettes and sometimes candy and soft drinks in a manner which established a friendly relaxed atmosphere. This was important as it immediately distinguished the interviewers from government "interrogators" who rarely allowed such amenities and often treated POWs and ralliers roughly. Most interviewees responded well to this approach although dedicated members of the PLA often refused American cigarettes but accepted Vietnamese brands.

As the interviews progressed, the respondents usually became more relaxed and more communicative. This was primarily because of the nature of the questions, which differed significantly from those asked by the military and police. The attitude of the interviewer was also important. As this is indicated in the Vietnamese language by the form of the promoun used, the interviewers always used respectful forms in their discussions with respondents and when referring to the PLA and the front. The interviewers never condemned past activities of the respondent and listened respectfully to interviewee responses. By 1965, most interviews were tape-recorded and supplemented by notes, which recorded nonverbal responses such as degree of relaxation and other such general impressions.

Once the actual interview started, the interviewer, knowing the questionnaire intimately, probed for those areas in which the respondent was most knowledgeable and developed them thoroughly. If the interviewer concluded the respondent's useful knowledge was exhausted, or that the respondent was being uncooperative, he ended the interview. For these reasons the length and substance of the interview reports vary widely.

THE VALIDITY AND RELIABILITY OF THE INTERVIEWS

A related question was whether the respondents told the truth. While there was usually no means to allow the interviewer to directly challenge statements by the respondent, the interviewers made every attempt to gain valid information. The best means to achieve this was through complete familiarity with the subject matter. Because more than 90 percent of all the interviews were produced by vastly experienced interviewers, evasive and untruthful answers were easily detected. The interviewers were also alert for inconsistencies and returned to suspect subject areas to resolve them. In other cases, the interviewer told the respondent that his replies did not sound like the truth and reminded him that he was not being forced to answer. Such cases sometimes led to acceptable answers and other times to the termination of the interview.

Generally, the status of the respondents, POW or defector, made little difference when responses were given. The expectation that defectors would be critical of the PLA and the front was not evident in the interviews. One interviewer stated:

> One of the objections (to the interviews) that might be raised. . .could be that defectors, since they had left the National Liberation Front, must all have been unsympathetic to the Front and favorably disposed toward the Saigon government. The truth, however, is that almost all the defectors took this step for personal reasons—very few did so for ideological reasons—and not because they supported the government. Most of them still thought that in many ways the Front was better than the government. . . .[12]

THE INTERVIEW REPORTS

Upon returning to Saigon from a field trip, the interviewer reconstructed the interview from his notes and tape recording. While small talk could be omitted from his report, he was carefully instructed to include anything that might possibly contribute to understanding the PLA and the front. In most interviews, both the question and response were written out in the report. First drafts of the reports were usually written in Vietnamese or French and then given to a translator for translation into English. The draft reports were then reviewed by the team leader or field director for editing and typed in final form.

The opportunities for error were plentiful. Different styles of writing by interviewers, incomplete notes, inaccuracies in translation, inaudible tapes, and other such sources of inaccuracies no doubt were reflected in the interview reports. However, as Davison observed, there were no cases in which interview reports were intentionally changed or relevant information left out.[13] Although no systematic procedures were used to check the quality of interview reports, checks were made from time to time. Comparisons were made between the final reports and portions of the interview tapes. External checks were made as well. Captured documents from other sources often dealt with matters similar to those related in interview reports. In this manner the general condition of PLA units as well as social and political conditions in certain areas reported in the interviews could often be corroborated. Thus, Davison notes that, while minor inaccuracies were probably evident in a number of areas, the interview reports are generally valid and reliable.

NOTES

1. The background information provided in this appendix draws on discussion with Konrad Kellen, project leader for the K-Series interviews, and on an interview user's guide. See W. Phillips Davison, *User's Guide to the Rand Interviews in Vietnam* (Santa Monica, Calif.: RAND Corporation, 1972).

2. Most of these interviews are available to the public as they were originally transcribed from the interviewers' notes and recordings. Only information that might allow a reader to identify the respondent has been deleted. This sanitizing process was taken to prevent possible reprisals against those who cooperated by responding to the questionnaire.

3. See Appendix B for the K-Series questionnaire. These interviews focused primarily on the motivation and morale of the individual PLA soldier. Studies based on interviews by the Elliotts, Zasloff, Goure, and others were generally narrowly focused on subjects such as interviewee attitudes toward bombing halts, the Tet talks, reasons for joining the PLA, impact of herbicide operations, party activities and organization, etc. No study addressed motivation and control of the PLA soldier within the overall context of PLA organization, policies, and relationships.

Studies such as those by Pike and Conley broach this subject but generally are cast so broadly that their treatment of the PLA is not as comprehensive and detailed as is attempted by the present study.

4. Davison, *User's Guide*, p. 5.

5. Ibid.

6. Ibid., pp. 6-7. Analysts who were familiar with the Vietnamese culture and language examined the overall interview system and were satisfied that the answers of the respondents were being accurately transcribed.

7. Ibid., p. 17.

8. Ibid., p. 18.

9. Ibid., pp. 26-27.

10. Ibid., p. 29.

11. Ibid., pp. 32-33.

12. Ibid., p. 36.

13. Ibid., p. 41.

Appendix B

QUESTIONNAIRE FOR MAIN FORCE PRISONERS AND RALLIERS[1]

I. Background

1. How long were you in your last unit (what kind of unit)?
   a. What other units have you served in?
2. (If cadre) When were you appointed a cadre? When were you appointed to the last position you held? What was your last position?

II. Recent Field Experiences

3. How many military operations did your unit (squad, platoon, company, battalion or regiment—find out which) go on in the last month before you were captured or rallied?
   a. How many during the six months before your capture or rally?
   b. How many of those operations did you participate in (in the last month? in the last six months?)?
   c. (If reply indicates very few operations by unit) Please explain why your unit engaged in so little combat.
   d. (If reply indicates frequent operations by unit) Please explain why so many operations.
4. How many times has your unit (define meaning of "unit") been attacked in the last month before your capture or rally by: aircraft, ground units, and artillery? (If ground units, what kind of ground units?)
   a. How many times has your unit been attacked in the last six months before your capture or rally?

b. Which attacks caused your unit the most damage and losses? How many lost? When did it happen?
c. How many losses did your unit (squad, platoon, company, or battalion) sustain in the six months before your capture or rally?

5. In the battles your unit participated in in the last six months, how many did your unit win, and how many times was it defeated?
   a. Please explain why your unit won and lost these battles.
   b. Has your unit been as successful, more successful, or less successful in battles in the past six months than in the time before that? Explain why.
6. How was the health of the men in your unit?
   a. How many in the past six months were usually too sick or weak to fight?
   b. Has the health of the men in your unit changed in the six months before your capture or rally as compared with previous times? Explain.
   c. Has the medical assistance given to the men in your unit in the six months before your capture or rally improved, been the same, or become worse as compared to previous times?
7. What were the greatest hardships you and your unit suffered in the six months before your capture or rally?
   a. Have the food rations issued to your unit been the same or different in these six months as compared to previous times? Describe and explain any change.
   b. How often did your unit move in the past six months? (If unit moved frequently) Why did your unit move so often? Has your unit moved more or less often than before? Explain.
   c. How did the moves affect the fighters? What did they say? What did the cadres say?
   d. In the past six months, did your unit usually stay in people's houses, huts, caves, or in the open? How did your unit usually camp before that time?
8. Describe the circumstances of your capture (or defection).
   a. If capture occurred in combat, how do you think you performed during this last battle?
   b. As best as possible, describe your feelings during the battle.
   c. Did you feel differently than in previous battles?
   d. Do you think you could have avoided capture? Explain.

III. Attitude Towards Unit Leadership

9. Describe the cadres in your unit (or squad, platoon, company, and battalion). What kind of persons were they?
   a. How much experience have they had?
   b. What did you think of them? What did the other men say?
   c. Did you have confidence in the unit's leaders? What did the other men say? Explain.
   d. In the six months before your capture or rally did you have as much, more, or less confidence in your unit's leaders than before? Explain why. What did the other men say?
10. Who was in charge of political education and morale problems in your unit? Did your unit have a "political cadre"? If so, what sort of person was he?
   a. What did the "political cadre" do? What were his responsibilities?
   b. Do you think his mission was an important one? Explain.
   c. What did you think of the political cadre? How did the other men feel towards him?
11. Did the "political cadre" in your unit do his work well? Elaborate.
   a. What did he do especially well? Was that good for you and the unit? Explain.
   b. What were his chief faults?
   c. What training and experience did he have? Was he qualified for his work?
   d. Did his behavior, attitudes, and performance change over time?
12. Was the "political cadre" (and the other cadres and Party members) successful in influencing the spirit and behavior of the men in the unit? How successful was he? Explain.
   a. How successful was he in bringing about good comradely relations and cooperation among the cadres in the unit between the cadres and fighters? Explain.
   b. What other cadres helped mobilize the spirit and watched over the correct behavior of the men? What did they do?
   c. Do you think that the cadre were sincere and told the fighters the truth? Did you believe in them? Why?
   d. In the last six months before your capture or rally, had you changed your opinion about the truthfulness of what the cadre said? If so, why?
13. If the fighters had problems of any kind, with whom could they discuss them? Elaborate.

a. What types of problems did the fighters feel free to discuss with the unit leaders (or political cadre)? What happened when they did so?
b. Were there any problems they avoided discussing? With the cadres? With other fighters? Elaborate.
c. Do you think the leaders of your unit had the right attitude towards the men's problems? Has their attitude changed in recent times? If so, why?

IV. Changes in Composition of Unit

14. What were the chief changes in the composition of your battalion, company, platoon, squad in the last 3-6 months before your capture or rally? Why did these changes take place?
15. (If subject mentions "changes in cadres") Did the new cadres come from some other unit, or were they promoted from within the company?
    a. What did you and the men think about the new cadres? Were they as good as the earlier cadres? Not as good? Explain.
16. (If subject mentions "new conscripts or new rank and file soldiers transferred to his unit from some other unit"):
    a. Where did they mostly come from (what units or areas)?
    b. Were these new soldiers in good physical condition? Well trained? Of high fighting spirit?
    c. Did the new soldiers fit into your unit and get along with the other men well? What did the other, older soldiers think of them?
17. Has your company had enough experienced platoon and squad leaders in recent months? How did they treat the men?
    a. Have the company or battalion cadres shown concern in any way about the quality and performance of the platoon and squad leaders? Elaborate.
18. How many Party members were there in your unit? Has this number increased, decreased, or remained the same while you were with the unit?
    a. How many regroupees were there in your unit? Has this number changed?
    b. What were the functions of these Party members and regroupees?

V. Surveillance, Discipline, Control

19. What attitude and behavior on the part of the men in your unit would be described as representing good morale and fighting spirit?

a. What did the cadres say the attitude and behavior of the fighters and cadres should be?
b. What were the most important indications that the morale and fighting spirit of the unit was good?

20. How did the cadre find out about the state of the morale, fighting spirit, and attitudes of the fighters?
 a. What signs did they look for?
 b. Who was responsible for keeping the company or battalion commanders informed about the morale and fighting spirit of the men in your unit?
 c. Do you think the commanders were always correctly informed about the morale or fighting spirit of the men? Explain.
21. Were the men always watched over by someone for signs of good and bad morale?
 a. How was the watching of the men organized?
 b. Did the men in the unit always know that they were being watched?
 c. Did the fighters in the unit always know who watched them?
 d. How did the fighters in the unit react to the knowledge that they were being watched for signs of bad morale, incorrect thoughts, poor performance of duties?
22. When you or the men in your unit were depressed, discouraged, or had complaints, did you ever talk to others about such things?
 a. To whom did you talk when you felt like that?
 b. When and under what circumstances did you talk like that?
 c. Did the men in the unit trust each other? Explain.
23. Did the fighters feel free to express their discontent, or to show that they were discouraged or depressed in front of squad, platoon, or company cadres?
 a. If so, what caused them most often to be so discouraged, depressed, or discontented as to show it to the cadres?
 b. Why did they show it to the cadres and when did they do so?
 c. In the six months before your capture or rally, had the fighters shown discouragement or discontent more, the same, or less frequently to the cadres than before? Elaborate.
24. Could you generally tell who in your unit had really good morale, and who merely pretended to? How could you tell?
25. Were there times when fighters worked or fought without cadres being there to watch over them or to lead them?
 a. If so, did the men do their work or fight well, or did they behave less well when not led by cadres? Explain why.

b. When men were separated from the unit during combat, or away from it while on a mission, did they do their best to rejoin the unit or obey their orders, or did they take advantage of being alone?

26. What do you think of the three-man cell system?
    a. Did it help the men's fighting spirit?
    b. Did it help the men in combat?
    c. Did it help the men to do their work when not in combat?
    d. Did it help you? If so, in what way?
    e. Was your best friend a member of your cell?
27. Were there ever any serious arguments between the fighters in your unit? Elaborate.
    a. Or between the fighters and the cadres? Elaborate.
    b. Or between the cadres in your unit? Elaborate.
    c. What were the most frequent causes of these arguments?
    d. In the past six months, have there been more or fewer arguments between the fighters and the cadres in your unit than before? If any changes have occurred, explain why.
28. Were you ever criticized in a criticism session or did you ever engage in self-criticism? If yes, how often, and what were the reasons?
    a. How did you feel about it?
    b. Do you think the criticism and self-criticism is good or bad? Do you think it is fair? Explain.
    c. What did the other men say?
    d. In your opinion, what influence did the criticism session (or the fear of being criticized) have on the behavior, fighting spirit, and performance of your unit?
    e. Did the criticism sessions correct and improve bad morale and incorrect thoughts? Explain.
    f. Did the criticism sessions make the men angry or discouraged?
29. In the last six months before your capture or rally, has the frequency of criticism sessions increased, remained the same, or decreased as compared to previous times?
    a. What were the most frequent subjects discussed at the criticism sessions? Have they changed in the past six months compared to previous times?
30. Were the demands the cadres and the leaders made on the fighters in your unit reasonable?
    a. Were the fighters willing to obey orders? Were they enthusiastic?

b. Were the fighters willing to obey even if they thought the orders were unreasonable? Give examples.
c. Were there cases where the fighters refused to carry out any order? If yes, give examples. What happened to the fighters who refused to carry out orders?
d. Have there been any cases where lower cadres or fighters have tried to have some orders changed? If yes, what kind of orders? What happened?
e. Were the fighters given a chance to discuss and criticize a plan of operation before the operation? Give details.

31. (If the interviewee is a cadre, ask "Did you always agree personally. . . .") Do you know whether the squad and platoon cadres always agreed personally with the orders from above? How do you know?
   a. Did they sometimes let the fighters know that they thought the orders were unreasonable, but had to be carried out anyway? If yes, what did they say?
   b. Did the squad or platoon cadres ever deliberately not carry out an order from above? Elaborate.
   c. Have there been any changes in the attitude of the squad and platoon cadres in the way they carry out orders or in not carrying out orders? Elaborate.
32. How have the cadres in your unit attempted to discourage or prevent the fighters from deserting home? Rallying? Surrendering in battle? Elaborate.
   a. Have there been any changes recently in the way the cadres have been doing it?
   b. What effect did these efforts have on the fighters?
33. Have there been any attempts at desertion home or rally to the government in your unit (squad, platoon, company) in the last month before your capture or rally? If so, how many? What sort of person? How many deserted and how many rallied?
   a. How many and what kind of persons deserted home, rallied, or let themselves be captured in the last six months before your capture or rally?
34. Why did the men rally or desert?
   a. Have the reasons for rallying or deserting changed in the last six months?
   b. What made the men decide to desert home, rather than rally, or to rally rather than desert home?
   c. Did the men in your unit talk among themselves about desertion or rallying? What did they say?

35. What were the fighters most afraid of in connection with deserting or rallying?
36. (If interviewee is a rallier) Why did you rally? Did you discuss your intention to rally with anyone before leaving your unit? If so, who was that person?
37. Did some of the fighters ask to be discharged and allowed to go home? How many asked to do so in the last three months before your capture or rally? What happened to them?
38. Did the cadres or the fighters try to avoid unpleasant or dangerous missions? What did they do? How many did so?
    a. Was there any change in that respect during the past six months before your capture or rally?
39. How would you describe the morale of your unit (good, average, poor, getting better or getting poorer, unchanged)?
40. Do you think that in the last six months the cadres were as effective or less effective than before in strengthening the fighting spirit of the men in your unit? Explain.
41. Have you and your comrades heard or seen GVN propaganda during the last six months?
    a. What propaganda from the GVN have you seen?
    b. What did you think of it?
    c. Do you think GVN propaganda during the last six months had more, less, or the same influence on the attitudes of the men in your unit than previously?
    d. What did the cadres do to reduce the influence GVN propaganda had on the attitudes of the men in your unit?

VI. Personal Experiences

42. What kept you fighting as long as you did?
    a. What kept the men in your unit fighting as long and as well as they did?
    b. Do you think that in the last six months the men in your unit fought as well as before? Explain.
43. When you were exhausted from an engagement or heavy exertions, with what kind of rest and relaxation were you provided?
    a. How did you personally spend such periods of rest?
    b. Were the men given more or less rest as time went on? Explain why.
    c. When a man was sick or wounded, where would he be given a chance to recuperate?
    d. Did you have any complaints about the whole system of rest and recuperation from fatigue and military action?

e. Was it discussed among the men and what was said?
f. Did you notice any changes with regard to these activities during the past year?

44. Did you have home leave?
 a. What was your unit's policy on home leave? Had it changed in the last year before your capture or rally?
 b. How did you feel about your unit's policy on home leave?
 c. Did you have home leave as often as you hoped? If not, how did you feel about not having enough home leave?
 d. What did the other fighters say?
45. Did you know how your family got along in your absence? How did you know about it?
 a. Did your family suffer any hardships in your absence? If yes, how did you feel about it?
46. When you went on an operation with your unit, did you feel that you were well prepared for it? Properly trained? Sufficiently well armed? Elaborate.
 a. Did you feel that your unit made sufficient efforts to remove and treat the wounded fighters and the dead?
 b. What did the other men say about the conditions of their families?
 c. Did you know what taxes your family paid to the Front during the past year? Do you think these taxes were reasonable?
 d. Do you think your family received help from the village Front organization while you were away? What did the other fighters say about the help their families received from Front organizations?
47. How did you feel with respect to engaging in battles: Did you feel that you became more experienced and less afraid as you gained combat experience, or did you feel that you became more frightened and less willing to fight?
 a. Do you think it becomes easier or harder for a fighter to fight when he stays in the war for a long time?
 b. Did you think your chances of surviving became better or poorer the longer you fought? The longer the war went on?
 c. What did the other men say?
48. When you were in combat how well did you fulfill your mission? Give details.
 a. Do you think you could have fought harder and more effectively? If so, why didn't you?
 b. Do you think the other men in your unit fought as hard as they could? Explain.

c. How did you personally feel when you were in combat?
d. Speaking frankly, did you ever engage in what your cadre called malingering while in combat? Give details.

49. Tell me about your experiences with villagers while you were with the Front. How often did you visit villages or talk with villagers?
    a. Did you have more, the same, or less contact with the villagers during the six months before your capture or rally than earlier?
    b. Were the villagers glad to see your unit and other Front and NVA units or were they afraid of them? Explain why.
    c. What was the villagers' attitude toward the Front civilian cadres and guerrillas?
    d. Did you notice any changes in the villagers' attitudes toward you and your unit during the time you were in the Front? During the last six months? What were the changes and what caused them?
50. Do you think the attitude and fighting spirit of the men in your unit was more or less influenced by the attitudes of the villagers toward the Front during the last six months?
    a. Did any changes in the villagers' attitudes make you feel differently about the war? In what way.

VII. Political Posture

51. Would you describe yourself as being a Communist? A democrat? A socialist? A revolutionary? Something else?
    a. How would you describe most of the others in your unit? The fighters? The cadres?
52. What do you think would be the best political system for South Vietnam to develop?
53. What did the cadres say about who started and is responsible for the war?
    a. What is your opinion on this?
    b. Who do you think is responsible for the war continuing for so long?
    c. Do you think the North was right in sending regroupees to South Vietnam to fight?
    d. Do you think the North was right in sending Northern soldiers to fight in South Vietnam? Do you think the North had the right to send these soldiers?
    e. Do you think the United States had the right to send its troops to South Vietnam?

f. Do you think the United States had the right to bomb North Vietnam? Do you think it was justified in doing so?
g. Do you think the Americans and the GVN forces are justified in bombing and shelling Front-controlled hamlets or contested areas?

VIII. Outcome of War

54. In the last month before your capture or rally, how did you think the war was going in your area of operation? Explain.
  a. Had your view of how the war was going changed during the last year before your capture or rally? Explain.
  b. (For captives) What did you think about continuing to fight? How long did you think you would have been willing to go on fighting?
55. In the six months before your capture or rally, did the cadres talk more than before about the way the war was going? If so, what did they say?
  a. Did you hear about the way the war was going from other sources (radio, leaflets, newspapers, comrades, or civilians)?
  b. Which sources did you hear more from in the last six months than before? Explain.
  c. Which information did you believe most? Why? What did your comrades say?
56. By what signs or events did you judge how the war was going in your area?
  a. What signs do you think were most important in showing who may win the war?
  b. What signs did your comrades talk about?
  c. What signs did the cadre talk about?
57. How do you think the war will end?
  a. Will the war be long or short? Explain.
  b. Will it end by negotiations or a military victory?
  c. Or will it never end? Explain.
58. What effect do you think the growing numbers of Americans and other foreign troops are having on the outcome of the war?
59. In your opinion, why do so many people in South Vietnam fight for and support the GVN?
  a. Do you think that support for the GVN among the people is becoming stronger or weaker?
  b. What signs or events have led you to believe that support for the GVN among the people is becoming stronger or weaker?
60. In your opinion, why do so many people in South Vietnam fight

for and support the NFL?

a. Do you think that support for the NLF among the people is becoming stronger or weaker?

b. What signs or events have led you to believe that support for the NLF among the people is becoming stronger or weaker?

61. If the GVN were to win the war, what do you think would happen to those who fight for the Front?

a. What do you think would happen to you and your family?

b. Your country?

62. If the Front should win the war, what do you think would happen?

a. What do you think would happen to you and your family?

b. Your country?

63. In the last months before your capture or rally, how did you feel about the importance of the Front's aims and of achieving them?

a. Did you think they were important enough for you to go on suffering hardships and risking being killed?

b. Did you think they were important enough to be worth the suffering of the people in a long war?

c. Did you think your children should go on fighting this war if this were necessary for the Front to win?

IX. Reasons for Joining (and Leaving) VC or NVA

64. Why did you join the VC (or NVA)?

a. Did you join of your own free will or were you coerced?

b. If of your own free will, what were your principal reasons?

c. If coerced, give details.

65. If of your own free will, did you find that the VC (or NVA) came up to your hopes and expectations? Surpassed them? Fell short of them? Elaborate.

a. Were you disappointed with the VC (or NVA)? When and why? Why did you leave?

66. Did you come to feel that you had made a mistake? Did you regret having joined? When and why?

67. If you turned away from the VC (or NVA), do you think you would have turned away from them also if they had won more battles and defeated the Americans?

a. Did your recent combat experience help shape your decision to leave?

b. What really made you turn away from them?

68. (In case of POW): How do you feel about the VC (or NVA) now? If you had the past to live over again, would you join again? Why? If not, why not?

NOTE

1. There was no standard interview format. Interviews were open-ended in order to allow interviewers to develop promising responses.

Bibliography

BOOKS

Almond, Gabriel A., and Verba, S. *The Civic Culture.* Boston: Little, Brown, 1965.

Bell, D. B., and Houston, Thomas, Jr. *The Vietnam Era Deserter.* Alexandria, Va., Army Research Institute, Report 76-6, 1976.

Berman, Paul. "The Liberation Armed Forces of the NLF, Compliance and Cohesion in a Revolutionary Army." Ph.D. dissertation, MIT, 1970.

Biesanz, John, and Biesanz, Marvis. *Modern Society.* Englewood Cliffs, N.J.: Prentice-Hall, 1968.

Conley, Michael C. *The Communist Insurgent Infrastructure in South Vietnam.* Washington, D.C.: Center for Research in Social Systems, American University, 1966.

Cooley, Charles H. *Social Organization.* New York: Charles Scribner's Sons, 1909.

Crawford, Ann. *Customs and Culture of Vietnam.* Tokyo: Tuttle, 1968.

Davison, W. Phillips. *User's Guide to the Rand Interviews in Vietnam.* Santa Monica, Calif.: RAND Corporation, March 1972.

Denton, Frank. *Volunteers for the Viet Cong,* RM-5647-ISA/ARPA. Santa Monica, Calif.: RAND Corporation, September 1968.

Dollard, John. *Fear in Battle.* New Haven, Conn.: Institute of Human Relations, Yale University, 1943.

Donnell, J.C.; Parker, G.J.; and Zasloff, J.J. *Viet Cong Motivation and Morale in 1964: A Preliminary Report,* RM-4507/3-ISA. Santa Monica, Calif.: RAND Corporation, March 1965.

Elliot, David W., and Elliot, Mai. *Part One: Unit Composition and Personnel,* RM-5848-ISA/ARPA. Santa Monica, Calif.: RAND Corporation, 1969.

——. *Part Two: Party Organization—Documents of an Elite VC Delta Unit*, RM-5849-ISA/ARPA. Santa Monica, Calif.: RAND Corporation, 1969.

——. *Part Three: Military Organization and Activities*, RM-5850-ISA/ARPA. Santa Monica, Calif.: RAND Corporation, 1969.

——. *Part Four: Political Indoctrination and Military Training*, RM-5851-ISA/ARPA. Santa Monica, Calif.: RAND Corporation, 1969.

——. *Part Five: Personal Letters*, RM-5852-ISA/ARPA. Santa Monica, Calif.: RAND Corporation, 1969.

Emerson, Rupert. *From Empire to Nation.* Boston: The Beacon Press, 1967.

Etzioni, Amitai. *Complex Organization.* New York: The Free Press, 1961.

FitzGerald, Frances. *Fire in the Lake.* New York: Random House, 1973.

French, J.R.P., Jr., and Raven, B. "The Bases of Social Power," in *The Study of Leadership.* West Point, N.Y.: USMA Printing Plant, 1971.

Gabriel, Richard A., and Savage, Paul L. *Crisis in Command: Mismanagement in the Army.* New York: Hill and Wang, 1978.

George, Alexander. *The Chinese Communist Army in Action.* New York: Columbia University Press, 1967.

Gettleman, Marvin E. *Viet Nam: History, Documents and Opinion.* New York: Fawcett, 1968.

Giap, Vo Nguyen. *Banner of People's War, The Party's Military Line.* New York: Praeger, 1970.

——. *Big Victory, Great Task.* New York: Praeger, 1968.

——. *Once Again We Will Win.* Hanoi: Foreign Languages Publishing House, 1966.

——. *People's War, People's Army.* Washington, D.C.: U. S. Government Printing Office, 1962.

Hammer, Ellen. *Vietnam, Yesterday and Today.* New York: Holt, Rinehart and Winston, 1966.

Hauser, William L. *America's Army in Crisis.* Baltimore: Johns Hopkins University Press, 1973.

Hickey, Gerald C. *Village in Vietnam.* New Haven, Conn.: Yale University Press, 1966.

Ho Chi Minh. *On Revolution.* New York: Praeger, 1967.

Huntington, Samuel P. *The Soldier and the State.* Cambridge, Mass.: Harvard University Press, 1967.

Jacobs, T. O. *Basic Problems in Small-Unit Leadership.* Fort Benning, Ga.: Hum RRO, 1965.

——. *Leadership and Exchange in Formal Organizations.* Alexandria, Va.: Human Resources Research Organization, 1971.

Janowitz, Morris, ed. *The New Military's Changing Patterns of Organization.* New York: Russell Sage Foundation, 1964.

———, and Little, R. *Sociology and the Military Establishment.* New York: Russell Sage Foundation, 1965.

Just, Ward. *Military Men.* New York: Avon Books, 1972.

Kellen, Konrad. *A View of the VC: Elements of Cohesion in the Enemy Camp in 1966-1967,* RM-5642-1-ISA/ARPA. Santa Monica, Calif.: RAND Corporation, November 1969.

———. *Conversations with NVA and VC Soldiers: A Study of Enemy Motivation and Morale,* RM-6131-1-ISA/ARPA. Santa Monica, Calif.: RAND Corporation, September 1970.

Kohn, Hans. *A History of Nationalism in the East.* London: Routledge, 1932.

———. *Nationalism: Its Meaning and History.* Princeton, N.J.: Van Nostrand Co., Inc., 1955, 1965.

Loory, Stuart H. *The Defeated: Inside America's Military Machine.* New York: Random House, 1973.

Mao Tse-Tung. *On Guerrilla Warfare.* New York: Praeger, 1967.

McCalister, John T. *Viet Nam: The Origins of Revolution.* New York: Knopf, 1969.

———, and Mus, Paul. *The Vietnamese and Their Revolution.* New York: Harper and Row, 1970.

McGarvey, P.J., ed. *Visions of Victory: Selected Vietnamese Communist Military Writings, 1964-1968.* Stanford, Calif.: Hoover Institution, 1969.

Marshall, S.L.A. *Men Against Fire.* New York: William Morrow, 1947.

Meyer, Frank S. *The Moulding of Communists.* New York: Harcourt, Brace, 1949.

Minh, Do Van. *Viet Nam: Where East and West Meet.* Milan: Casa Editrice, n.d.

Molnar, Andrew R. *Human Factors: Considerations of Undergrounds in Insurgencies.* Washington, D.C.: Special Operations Research Office, American University, 1965.

Moskos, Charles C., Jr. *The American Enlisted Man.* New York: Russell Sage Foundation, 1970.

———. "A Sociologist Appraises the GI," in *The Study of Leadership.* West Point, N.Y.: USMA Printing Plant, 1972.

Pike, Douglas. *Viet Cong, The Organization and Techniques.* Cambridge, Mass.: MIT Press, 1966.

Polsby, Nelson, et al. *Politics and Social Life.* Boston: Houghton Mifflin, 1963.

Roblant, Samuel. *The Israeli Soldier.* Cranbury, N.J.: Thomas Yoseloff, 1970.

Selznick, Philip. *The Organizational Weapon.* New York: McGraw-Hill, 1952.

Sharp, U.S.G., and Westmoreland, W.C. *Report on the War in Vietnam.* Washington, D.C.: U. S. Government Printing Office, 1968.

Stouffer, Samuel A., et al. *The American Soldier*, vol. 4. Princeton, N.J.: Princeton University Press, 1949.

Swearingen, R., and Rolph, H. *Communism in Vietnam: A Documentary Study.* Chicago, Ill.: American Bar Association, 1967.

Tanham, George K. *Communist Revolutionary Warfare.* New York: Praeger, 1967.

Tran Van Tung. *Les Hommes au-delà du Sud.* Paris: Editions de la Baconnière, 1957.

——. *Vietnam.* New York: Praeger, 1959.

Zasloff, J. J. *The Viet Minh Regroupees.* Santa Monica, Calif.: RAND Corporation, 1966.

——. *Origins of the Insurgency in South Vietnam, 1954-1960: The Role of the Southern Vietminh Cadres.* Santa Monica, Calif.: RAND Corporation, 1968.

——. *Political Motivation of the Viet Cong: The Political Regroupees,* RM-4703/2-ISA/ARPA. Santa Monica, Calif.: RAND Corporation, May 1968.

GOVERNMENT DOCUMENTS

Leadership for the 1970's. Carlisle Barracks, Pa.: U. S. Army War College, 1970.

Study on Military Professionalism. Carlisle Barracks, Pa.: U. S. Army War College, 1970.

VIET NAM DOCUMENTS AND RESEARCH NOTES (published as a research service by the U.S. Embassy in Saigon)

Document no. 15. "Troop Training and Combat Competition Campaign—An Emulation Plan."

Document no. 23. Pamerlee, Albert E. "Viet Cong Political Geography."

Document no. 34. "The Party in Command: Political Organization and the Viet Cong Armed Forces."

Document no. 40. Pamerlee, Albert E. "The Central Office of South Viet-Nam."

Documents no. 43-44. "Friction Between Northern and Southern Vietnamese: Directives Urge Standing Shoulder-to-Shoulder with our Kith and Kin Brothers."

Documents no. 58-59. "Self Sufficiency: A Duty of Cadre and Combatant."
Document no. 68. "Two Speeches by General Vo Nguyen Giap."
Document no. 70. Giap, Vo Nguyen. "The Party's Military Line Is the Ever-Victorious Banner of People's War in Our Country."
Document 71. Dung, Van Tien. "Under the Party's Banner, Viet-Nam's Military Art Has Constantly Developed and Triumphed."
Document no. 72. Hao, Song. "Party Leadership Is the Cause of the Growth and Victories of our Army."
Document no. 73. "Ideological Deficiencies and Lowered Combat Effectiveness."
Document no. 84. "A Brief Chronology of Momentous Facts and Events in the History of the Democratic Republic of Viet-Nam."
Document no. 87. Giap, Vo Nguyen. "Developing Vietnamese Military Science."
Document no. 102. "The PRPSVN—Part II: COSVN's Attempt to Revitalize the PRP."

PERIODICALS AND ARTICLES

George, Alexander L. "Primary Groups, Organization and Military Performance," in *The Study of Leadership.* West Point, N.Y.: USMA Printing Plant, 1972.
Henderson, William D. "An Integrated Concept of Leadership: A Functional Model," in *The Study of Leadership.* West Point, N.Y.: USMA Printing Plant, 1972.
Homans, George C. "The Small Warship." *American Sociological Review,* II, 1946.
Janis, Irving L. "Group Identification Under Conditions of External Danger," in *The Study of Leadership.* West Point, N.Y.: USMA Printing Plant, 1971.
Janowitz, Morris. "Hierarchy and Authority in the Military Establishments," in Etzioni, Amitai, *A Sociological Reader.* New York: Holt, 1969.
Jumper, Roy, and Normand, Marjorie Weiner. "Vietnam: The Historical Background," in Gettleman, Marvin E., *Viet Nam: History, Documents and Opinion.* New York: Fawcett, 1968.
Little, Roger W. "Buddy Relations and Combat Performance," in Janowitz, M., ed., *The New Military: Changing Patterns of Organization.* New York: Russell Sage Foundation, 1964.
——. "Finding a Buddy," in *The Study of Leadership.* West Point, N.Y.: USMA Printing Plant, 1972.

Moskos, Charles C., Jr. "The American Combat Soldier in Vietnam," *Journal of Social Issues*, vol. 31, no. 4 (1975).

Mus, Paul. "Vietnam: A Nation Off Balance," *Yale Review* 41 (1952).

Secord, Paul F., and Backman, Carl W. "Festinger's Theory of Cognitive Dissonance," in *The Study of Leadership*. West Point, N.Y.: USMA Printing Plant, 1971.

Shils, Edward A. "Primary Groups in the American Army," in Merton, R.K., and Lazarsfeld, P.F., *Studies in the Scope and Method of "The American Soldier."* New York: The Free Press, 1950.

——, and Janowitz, M. "Cohesion and Disintegration in the Wehrmacht in World War II," *Public Opinion Quarterly* 12 (2), 1948.

Turley, William S. "The Political Role and Development of the People's Army of Vietnam." Carbondale, Ill.: Department of Political Science, Southern Illinois, 1975 (presented at the American Political Science Association Convention, San Francisco, Calif.: Sept. 1975).

INTERVIEWS

The K-Series interviews (for Konrad Kellen) were quoted extensively. These interviews were conducted in Vietnam under the auspices of the RAND Corporation, Santa Monica, California, from 1965 through 1967. See Appendix A for a description of the interviews.

Index

About the Author

Lt. Colonel William Darryl Henderson is presently a battalion commander in the United States Army. He has served as a company commander in Vietnam and in the offices of the Secretary of Defense and the Chief of Staff of the U.S. Army. A Vietnam combat veteran with a Ph.D. in International Relations, he has taught military psychology and international relations at West Point. Colonel Henderson co-authored the *Handbook of World Conflict* and holds the Purple Heart.